Developing Numeracy
MENTAL MATHS

ACTIVITIES FOR THE DAILY MATHS LESSON

year
5

Hilary Koll and Steve Mills

A & C BLACK

Contents

Mental calculation strategies (x and ÷)

Answers

Reprinted 2007
First published 2004 by A & C Black Publishers Limited
38 Soho Square, London W1D 3HB
www.acblack.com

ISBN 978-0-7136-6914-5

The authors and publishers would like to thank Jane McNeill and Catherine Yemm for their advice in producing this series of books.

A CIP catalogue record for this book is available from the British Library.

Printed and bound in Great Britain by Cromwell Press Ltd, Trowbridge.

A & C Black uses paper produced with elemental chlorine-free pulp, harvested from managed sustainable forests.

Introduction

Developing Numeracy: Mental Maths is a series of seven photocopiable activity books designed to be used during the daily maths lesson. This book focuses on the skills and concepts for mental maths outlined in the National Numeracy Strategy *Framework for teaching mathematics* for Year 5. The activities are intended to be used in the time allocated to pupil activities; they aim to reinforce the knowledge and develop the facts, skills and understanding explored during the main part of the lesson. They provide practice and consolidation of the objectives contained in the framework document.

Mental Maths Year 5

To calculate mentally with confidence, it is necessary to understand the three main aspects of numeracy shown in the diagram below. These underpin the teaching of specific mental calculation strategies.

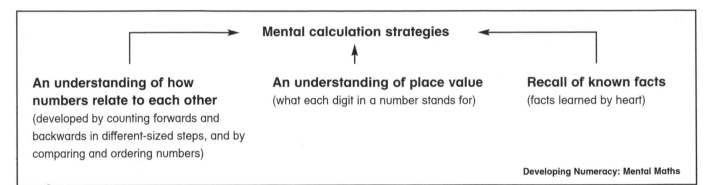

Mental calculation strategies

An understanding of how numbers relate to each other
(developed by counting forwards and backwards in different-sized steps, and by comparing and ordering numbers)

An understanding of place value
(what each digit in a number stands for)

Recall of known facts
(facts learned by heart)

Developing Numeracy: Mental Maths

Year 5 supports the teaching of mental maths by providing a series of activities which develop these essential skills. On the whole the activities are designed for children to work on independently, although this is not always possible and occasionally some children may need support.

Year 5 develops concepts and skills for the different aspects of numeracy in the following ways:

An understanding of how numbers relate to each other

- recognising and extending number sequences formed by counting from any number in steps of constant size, extending beyond zero when counting back;
- recognising multiples of 6, 7, 8 and 9, up to the 10th multiple, and finding all pairs of factors of any number up to 100.

An understanding of place value

- reading and writing whole numbers in figures and words, and knowing what each digit represents;
- multiplying and dividing any positive integer up to 10 000 by 10 or 100, and understanding the effect;
- using the vocabulary of comparing and ordering numbers, including symbols such as <, > and =, and giving one or more numbers lying between two given numbers;
- ordering integers less than one million.

Recall of known facts

Know by heart or derive quickly:

- decimals that total 1 or 10, all two-digit pairs that total 100, and all pairs of multiples of 50 with a total of 1000;
- all multiplication facts up to 10 × 10, and squares of numbers to at least 10 × 10;
- division facts corresponding to tables up to 10 × 10;
- doubles of all whole numbers 1 to 100, doubles of multiples of 10 to 1000 and doubles of multiples of 100 to 10 000, and the corresponding halves.

Mental calculation strategies

- finding a small difference by counting up through the next multiple of 10, 100 or 1000;
- partitioning when adding or multiplying;
- identifying near doubles, such as 1·5 + 1·6;
- adding or subtracting the nearest multiple of 10, then adjusting;
- adding several numbers and multiples of 10, and checking the sum of several numbers by adding in reverse order;
- using known number facts and place value for mental addition and subtraction;
- using doubling and halving, starting from known facts;
- using known facts and place value to multiply and divide;
- beginning to use brackets;
- using factors and closely related facts.

Quiz kids

- **Tick the correct answer for each question.**

Ellie **Jack** **Mira**

		Ellie	Jack	Mira
1.	$320 \div 10$	3200	320	32 ✓
2.	430×10	4300	43 000	43
3.	$760 \div 10$	7·6	7600	76
4.	57×10	5700	570	5·7
5.	$2100 \div 10$	21 000	210	21
6.	98×10	9800	98 000	980
7.	$9600 \div 10$	96	9·60	960
8.	$46 \div 10$	46·0	460	4·6
9.	$62·0 \div 10$	62	6·2	0·62

- **Who wins the quiz?** _____

Now try this!

- | Divide by 100 | **to convert these amounts.**

$450p = £\underline{\ 4.50\ }$ $620\,cm = $ _____ m $1130p = £$ _____

$6760p = £$ _____ $325\,cm = $ _____ m $1758p = £$ _____

Teachers' note It is important that the children appreciate that it is the digits that move to the left or right when multiplying or dividing by 10 or 100, and that zeros are used as place holders to indicate the columns that are empty.

Developing Numeracy Mental Maths Year 5 © A & C BLACK

Cross 'em out game

- **Play this game with a partner.**

☆ On another piece of paper, each player writes ten numbers between 10 000 and 999 999.

☆ Cut out the cards below. Place them face down.

☆ Take turns to pick up a card. If the description matches one of your numbers, keep the card and cross off the number. If not, turn it face down again. If the description matches several numbers, choose <u>one</u> to cross off.

☆ The winner is the first to cross off all ten numbers, or the player with the most numbers crossed off at the end of the game.

A number < 40 000	A number > 50 000	A number between 40 000 and 80 000
A number < 25 000	A number > 67 000	A number between 32 000 and 50 000
A number between 42 000 and 68 000	A number < 500 000	A number > 750 000
A number > 175 000	A number < 567 000	A number between 450 000 and 650 000
A number > 800 000	A number between 800 000 and 899 000	A number < 100 000
A number between 600 000 and 800 000	A number > 100 000	A number between 68 000 and 99 000
A number between 75 000 and 125 000	A number > 900 000	A number < 20 000
A number between 10 000 and 80 000	A number between 450 000 and 550 000	A number < 356 981

Teachers' note At the start of the lesson, revise the 'greater than' and 'less than' signs. Give pairs of five- or six-digit numbers and encourage the children to say all the numbers that lie between them.

**Developing Numeracy
Mental Maths Year 5
© A & C BLACK**

Going hot and cold

The temperature on the thermometer keeps rising and falling.

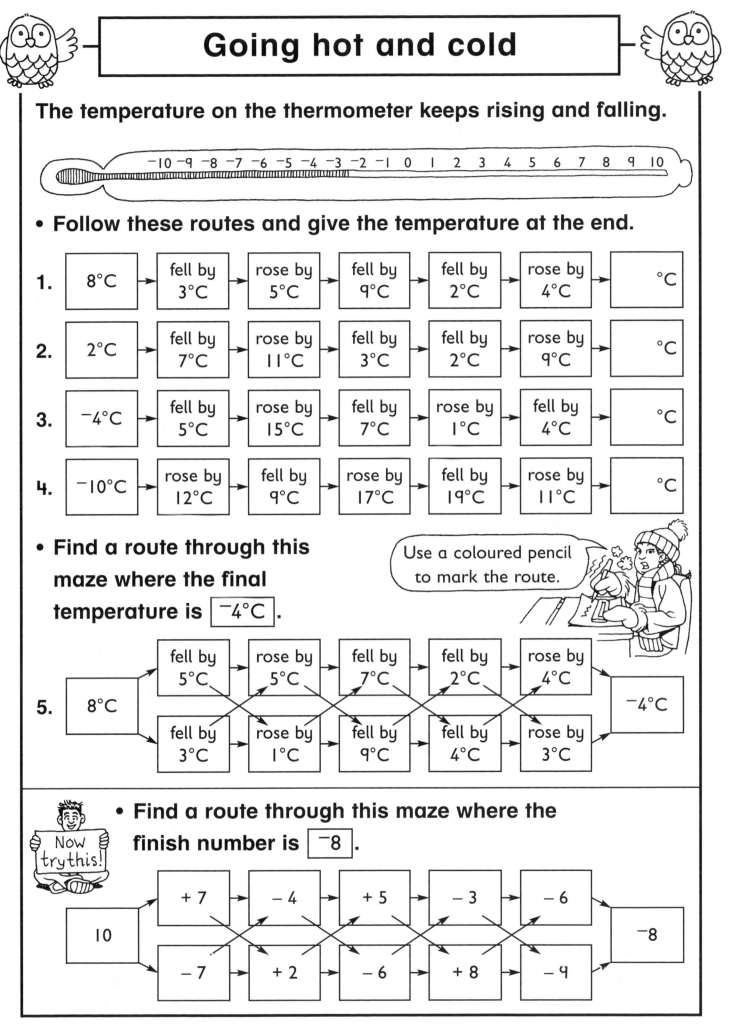

-10 -9 -8 -7 -6 -5 -4 -3 -2 -1 0 1 2 3 4 5 6 7 8 9 10

• Follow these routes and give the temperature at the end.

1. | 8°C | → | fell by 3°C | → | rose by 5°C | → | fell by 9°C | → | fell by 2°C | → | rose by 4°C | → | °C |

2. | 2°C | → | fell by 7°C | → | rose by 11°C | → | fell by 3°C | → | fell by 2°C | → | rose by 9°C | → | °C |

3. | ⁻4°C | → | fell by 5°C | → | rose by 15°C | → | fell by 7°C | → | rose by 1°C | → | fell by 4°C | → | °C |

4. | ⁻10°C | → | rose by 12°C | → | fell by 9°C | → | rose by 17°C | → | fell by 19°C | → | rose by 11°C | → | °C |

• **Find a route through this maze where the final temperature is** ⁻4°C .

Use a coloured pencil to mark the route.

5. | 8°C |

fell by 5°C → rose by 5°C → fell by 7°C → fell by 2°C → rose by 4°C → ⁻4°C

fell by 3°C → rose by 1°C → fell by 9°C → fell by 4°C → rose by 3°C

Now try this!

• **Find a route through this maze where the finish number is** ⁻8 .

| 10 |

+ 7 → − 4 → + 5 → − 3 → − 6 → ⁻8

− 7 → + 2 → − 6 → + 8 → − 9

Teachers' note Demonstrate using the number line to show the changes in temperature. Encourage the children to discuss other strategies for finding the new temperature, such as bridging through zero: for example, 'When counting back 7 from 3, I count back 3 first to get to zero. I know that 7 − 3 = 4, so 4 more takes me to ⁻4.' They will need to use trial and error to solve the mazes.

Developing Numeracy Mental Maths Year 5 © A & C BLACK

Great grids: 1

- **Complete the grid. Then write the sequence of numbers in the circles. Describe the sequence.**

> The numbers increase as you move down or to the right.

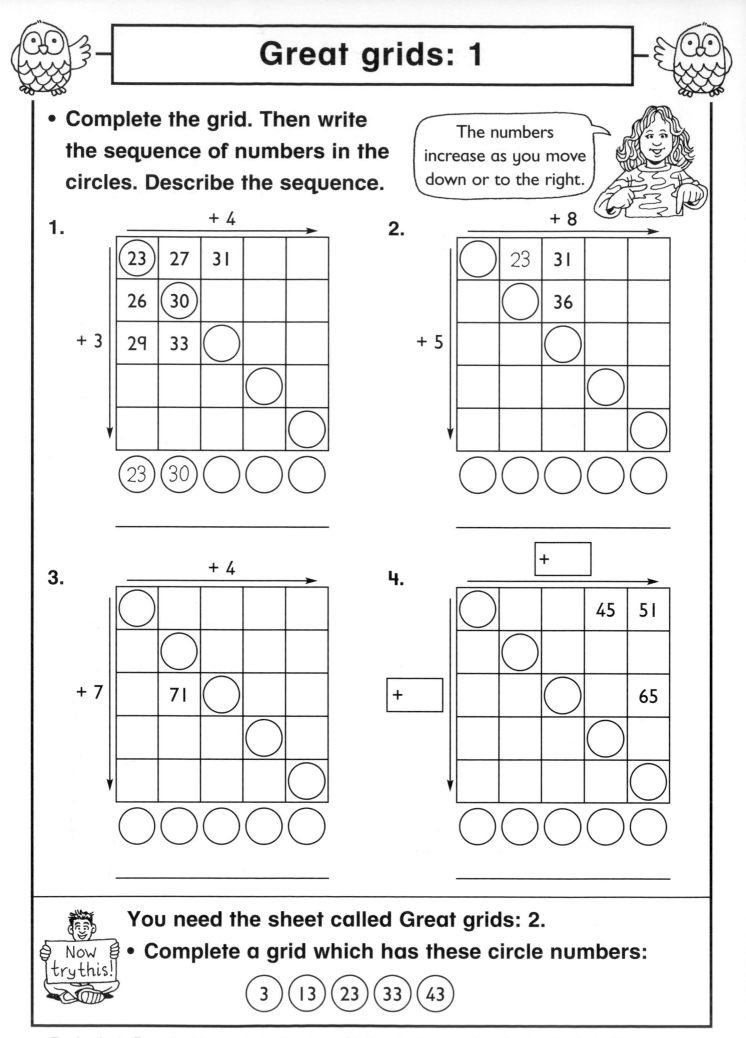

You need the sheet called **Great grids: 2.**

- **Complete a grid which has these circle numbers:**

(3) (13) (23) (33) (43)

Teachers' note Ensure the children understand how to complete the grids. Encourage discussion of ways to predict and find answers, using subtraction where necessary. Discuss the importance of checking mental calculations, working both across and down as a check. The children will need a copy of page 11 for the extension activity.

**Developing Numeracy
Mental Maths Year 5
© A & C BLACK**

Great grids: 2

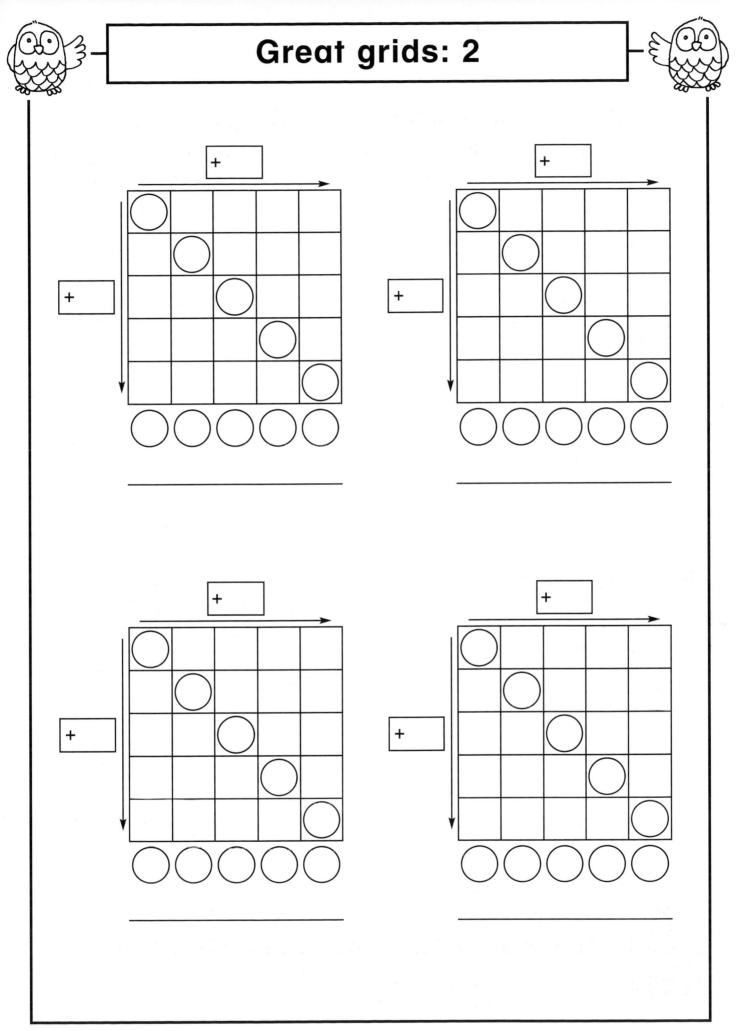

Teachers' note Use this when the children have completed the activity on page 10. The numbers could be filled in before photocopying or the children could investigate patterns using their own choice of numbers. Alternatively, they could be asked to investigate different ways of completing the grids for given circle numbers, such as (41) (49) (57) (65) (73).

Developing Numeracy Mental Maths Year 5 © A & C BLACK

11

Cross sequences

• **Find the circled number in the puzzle. Follow the instructions.**

① Count on in 21s	② Count on in 8s	③ Count back in 18s
④ Count back in 20s	⑤ Count back in 9s	⑥ Count on in 6s
⑦ Count on in 5s	⑧ Count on in 14s	⑨ Count back in 7s
⑩ Count back in 6s	⑪ Count on in 3s	⑫ Count on in 9s
⑬ Count back in 11s	⑭ Count back in 30s	⑮ Count on in 25s
⑯ Count on in 35s	⑰ Count back in 19s	⑱ Count on in 19s
⑲ Count back in 11s	⑳ Count back in 11s	㉑ Count back in 15s
㉒ Count back in 5s	㉓ Count back in 19s	㉔ Count back in 23s

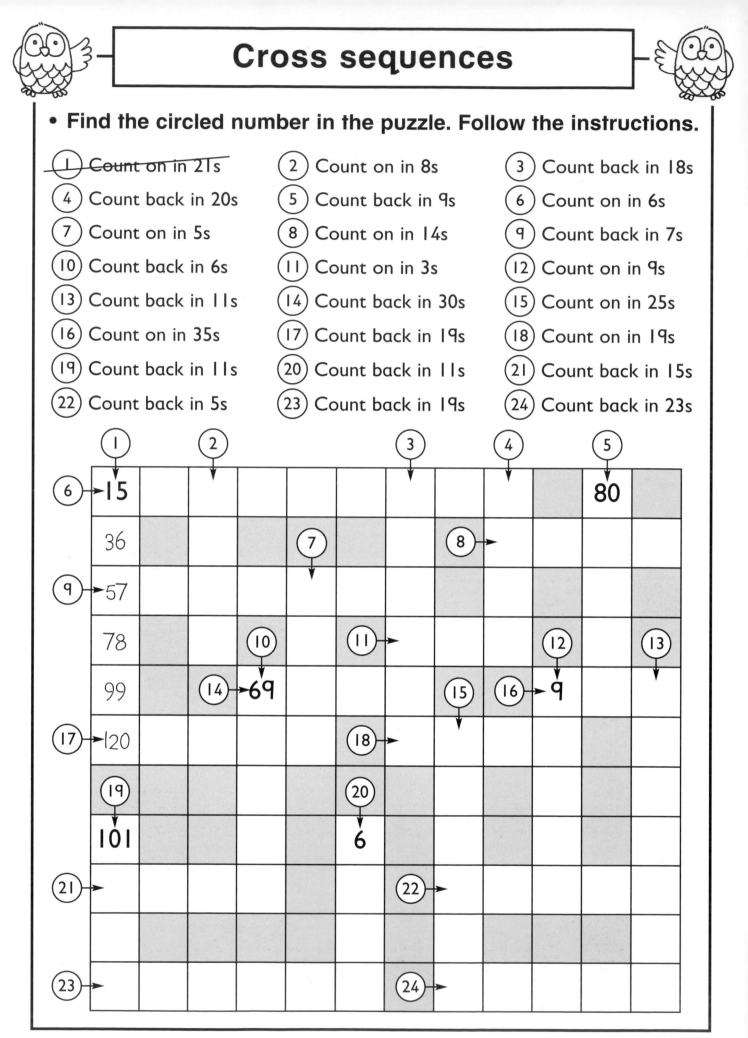

Teachers' note Revise counting on and back in equal steps, including counting on in steps of 11, 19, 21, 29, 25 and 50, and counting back beyond zero. The children will need to use negative numbers when completing this puzzle. Some children may need to use a number line and make jottings when bridging through zero.

**Developing Numeracy
Mental Maths Year 5
© A & C BLACK**

Factor tractors

• Cross off the numbers that do not divide exactly into the number on the wheel. Then write other ⬚factors⬚ of the number.

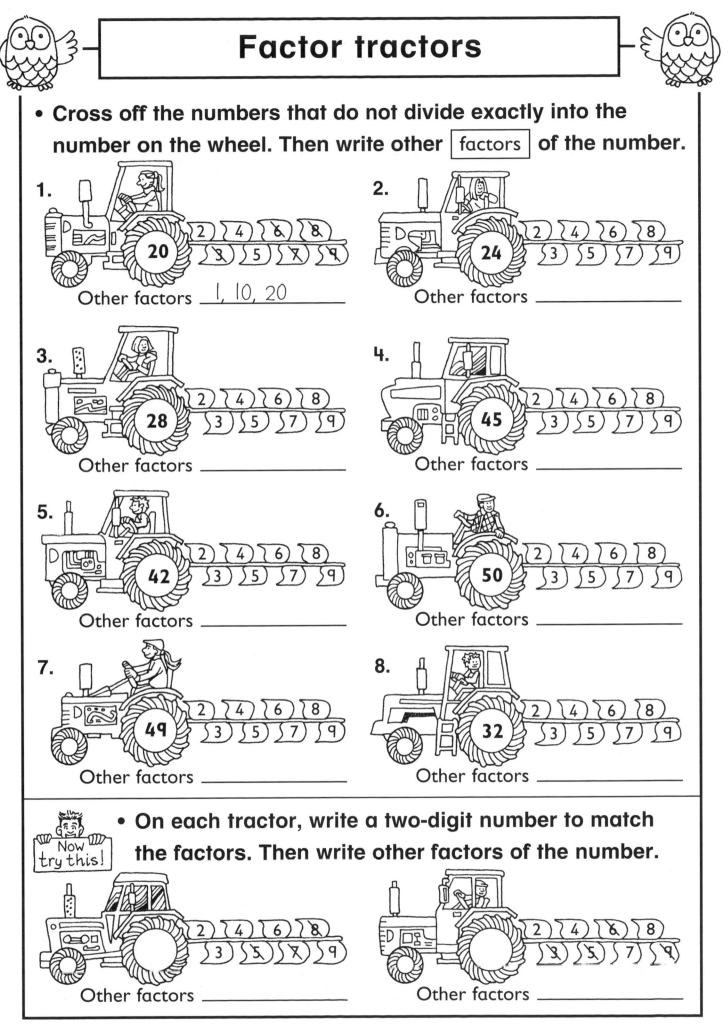

1. **20** · 2 4 ~~6~~ 8 ~~3~~ 5 ~~7~~ ~~9~~

Other factors ___1, 10, 20___

2. **24** · 2 4 6 8 3 5 7 9

Other factors _____

3. **28** · 2 4 6 8 3 5 7 9

Other factors _____

4. **45** · 2 4 6 8 3 5 7 9

Other factors _____

5. **42** · 2 4 6 8 3 5 7 9

Other factors _____

6. **50** · 2 4 6 8 3 5 7 9

Other factors _____

7. **49** · 2 4 6 8 3 5 7 9

Other factors _____

8. **32** · 2 4 6 8 3 5 7 9

Other factors _____

Now try this!

• On each tractor, write a two-digit number to match the factors. Then write other factors of the number.

2 4 6 ~~8~~ 3 ~~5~~ ~~7~~ 9

Other factors _____

2 4 ~~6~~ 8 ~~3~~ ~~5~~ 7 ~~9~~

Other factors _____

Teachers' note Revise the meaning of the term 'factor'. When the children are writing other factors, remind them that they can always write 1 and the number itself. To find others, they can divide the number by the factors they already have (for example, 20 has the factor 2, and 2 × 10 = 20, so 10 is also a factor of 20). Ask the children to explore other numbers in the same way.

**Developing Numeracy
Mental Maths Year 5
© A & C BLACK**

Dot your 'eyes'...

- **Each pair of decimals totals** $\boxed{1}$ **. Fill in the missing digits.**

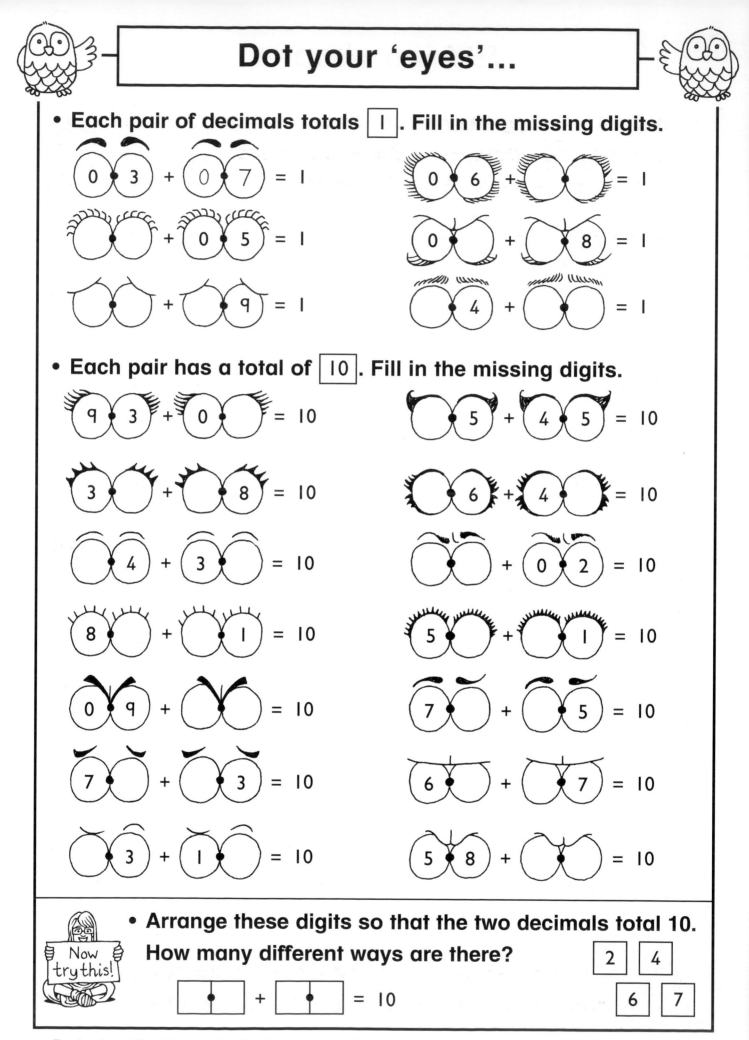

$(0 \cdot 3) + (0 \cdot 7) = 1$

$(\cdot) + (0 \cdot 5) = 1$

$(\cdot) + (\cdot 9) = 1$

$(0 \cdot 6) + (\cdot) = 1$

$(0 \cdot) + (\cdot 8) = 1$

$(\cdot 4) + (\cdot) = 1$

- **Each pair has a total of** $\boxed{10}$ **. Fill in the missing digits.**

$(9 \cdot 3) + (0 \cdot) = 10$

$(3 \cdot) + (\cdot 8) = 10$

$(\cdot 4) + (\cdot 3) = 10$

$(8 \cdot) + (\cdot 1) = 10$

$(0 \cdot 9) + (\cdot) = 10$

$(7 \cdot) + (\cdot 3) = 10$

$(\cdot 3) + (\cdot 1) = 10$

$(\cdot 5) + (4 \cdot 5) = 10$

$(\cdot 6) + (\cdot 4) = 10$

$(\cdot) + (0 \cdot 2) = 10$

$(5 \cdot) + (\cdot 1) = 10$

$(7 \cdot) + (\cdot 5) = 10$

$(6 \cdot) + (\cdot 7) = 10$

$(5 \cdot 8) + (\cdot) = 10$

- **Arrange these digits so that the two decimals total 10.**

How many different ways are there?

$\boxed{} \cdot \boxed{} + \boxed{} \cdot \boxed{} = 10$

$\boxed{2} \boxed{4}$

$\boxed{6} \boxed{7}$

Now try this!

Teachers' note The children could be timed to see how long it takes them to complete the main activity. They could be given the sheet at a later date and timed again to check for improvement. When finding decimals that total 10, encourage the children to use their knowledge of pairs of numbers with a total of 100.

**Developing Numeracy
Mental Maths Year 5
© A & C BLACK**

14

Hidden shapes

- On each square, join pairs of numbers with a total of $\boxed{100}$. Colour the shape inside the lines you have drawn. Write its name.

Use a ruler.

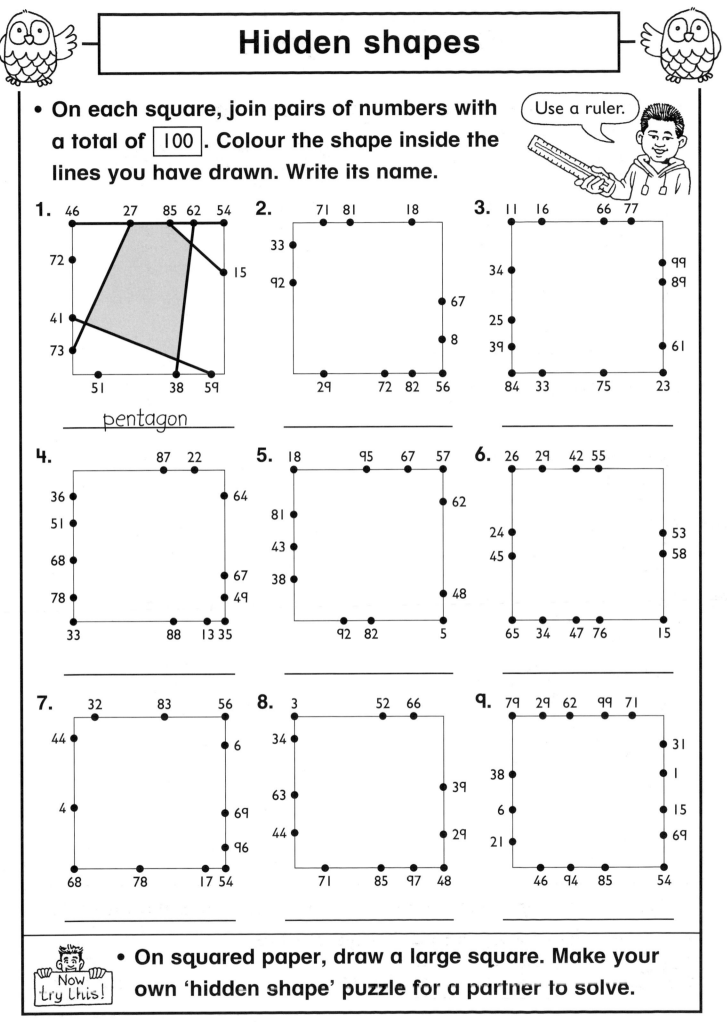

1. 46   27   85   62   54
72
15
41
73
51   38   59

pentagon

2. 71 81   18
33
92
67
8
29   72 82 56

3. 11 16   66 77
34
99
25
89
39
61
84 33   75   23

4.   87   22
36
64
51
68
78
67
49
33   88 13 35

5. 18   95   67   57
81
62
43
38
48
92 82   5

6. 26 29   42 55
24
53
45
58
65   34   47 76   15

7. 32   83   56
44
6
4
69
96
68   78   17 54

8. 3   52 66
34
63
39
44
29
71   85 97 48

9. 79 29 62 99 71
31
38
1
6
15
21
69
46 94 85   54

- On squared paper, draw a large square. Make your own 'hidden shape' puzzle for a partner to solve.

Teachers' note Revise the shape names 'triangle', 'square', 'rectangle', 'quadrilateral', 'pentagon', 'hexagon' and 'octagon'. The children could use a coloured pencil to draw the lines. Remind them that the units digits will add to make 10, so the tens must total 90 only. (Watch out for children making mistakes such as 72 + 38 = 100.) Provide squared paper for the extension activity.

**Developing Numeracy
Mental Maths Year 5
© A & C BLACK**

Mighty mice

Each mouse lifts a total mass of ⬚ 1 kg (1000 g).

- **Fill in the missing mass.**

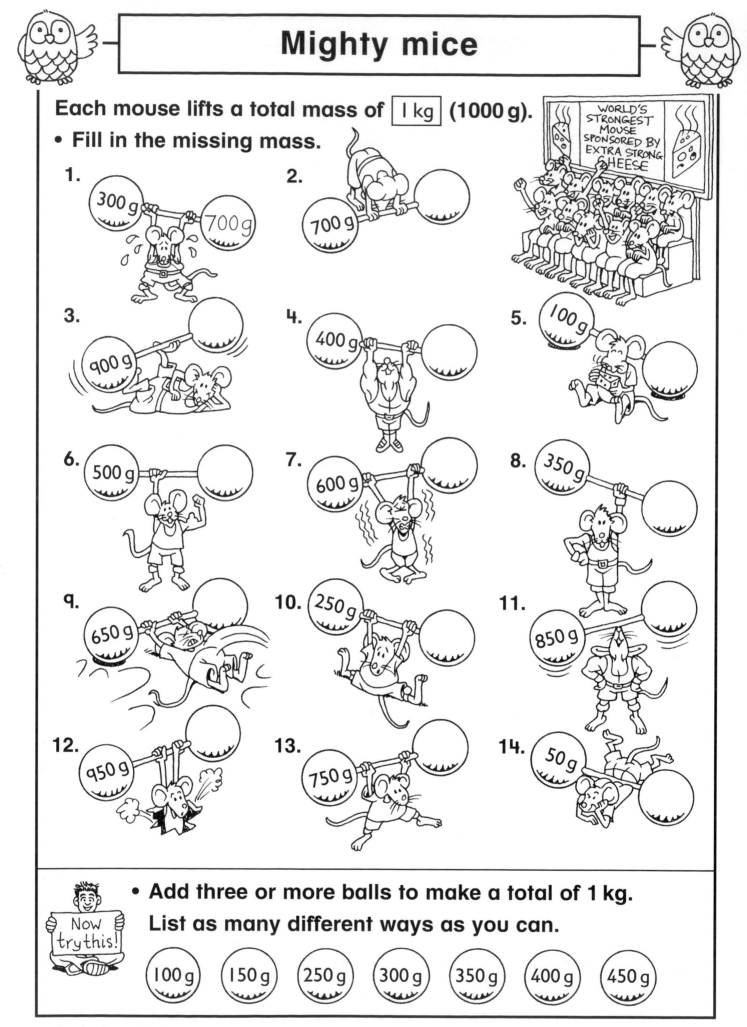

WORLD'S STRONGEST MOUSE SPONSORED BY EXTRA STRONG CHEESE

1. 300 g   700 g

2. 700 g

3. 900 g

4. 400 g

5. 100 g

6. 500 g

7. 600 g

8. 350 g

9. 650 g

10. 250 g

11. 850 g

12. 950 g

13. 750 g

14. 50 g

Now try this!
- **Add three or more balls to make a total of 1 kg.**
- **List as many different ways as you can.**

100 g   150 g   250 g   300 g   350 g   400 g   450 g

Teachers' note The children could be timed to encourage rapid recall of pairs of multiples of 50 with a total of 1000. The sheet could be altered to give practice in finding pairs of numbers with a total of 2 kg (2000 g). The values in question 1 would then be 300 g and 1700 g.

Developing Numeracy
Mental Maths Year 5
© A & C BLACK

Age old questions

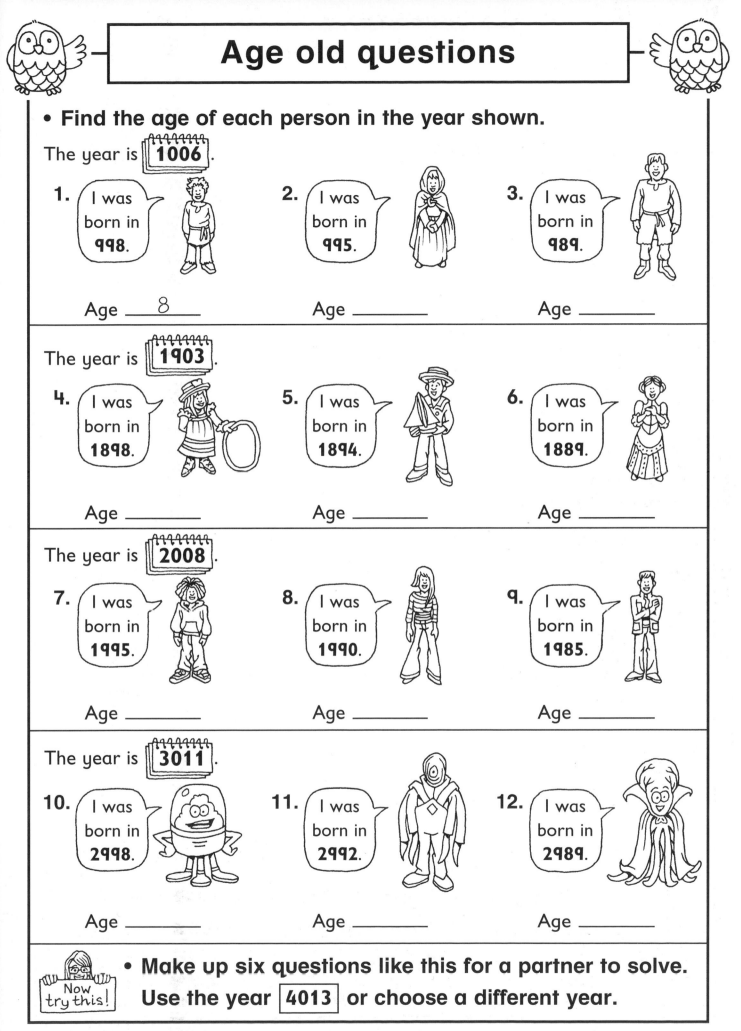

• **Find the age of each person in the year shown.**

The year is 1006.

1. I was born in **998**.
Age ___8___

2. I was born in **995**.
Age _____

3. I was born in **989**.
Age _____

The year is 1903.

4. I was born in **1898**.
Age _____

5. I was born in **1894**.
Age _____

6. I was born in **1889**.
Age _____

The year is 2008.

7. I was born in **1995**.
Age _____

8. I was born in **1990**.
Age _____

9. I was born in **1985**.
Age _____

The year is 3011.

10. I was born in **2998**.
Age _____

11. I was born in **2992**.
Age _____

12. I was born in **2989**.
Age _____

Now try this!

• **Make up six questions like this for a partner to solve.**
Use the year 4013 or choose a different year.

Teachers' note Remind the children that these questions can be written as subtractions: for example, 1006 – 998. Demonstrate counting on from the smaller number, using the multiple of 100 or 1000 as a stopover: for example, 'I know that 998 is 2 less than 1000 and 1006 is 6 more than 1000, so 1006 – 998 is 2 + 6 = 8.'

Developing Numeracy
Mental Maths Year 5
© A & C BLACK

Let's split!

- **Split the numbers into hundreds, tens and units to help you find the total.**

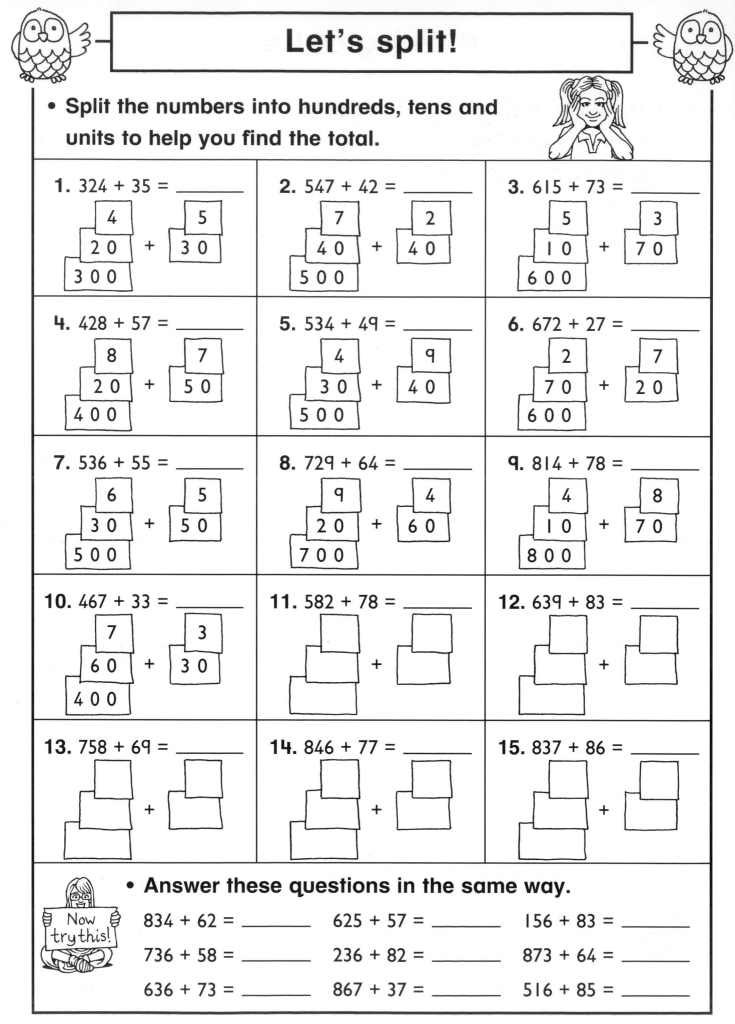

1. 324 + 35 = _____

4		5
2 0	+	3 0
3 0 0		

2. 547 + 42 = _____

7		2
4 0	+	4 0
5 0 0		

3. 615 + 73 = _____

5		3
1 0	+	7 0
6 0 0		

4. 428 + 57 = _____

8		7
2 0	+	5 0
4 0 0		

5. 534 + 49 = _____

4		9
3 0	+	4 0
5 0 0		

6. 672 + 27 = _____

2		7
7 0	+	2 0
6 0 0		

7. 536 + 55 = _____

6		5
3 0	+	5 0
5 0 0		

8. 729 + 64 = _____

9		4
2 0	+	6 0
7 0 0		

9. 814 + 78 = _____

4		8
1 0	+	7 0
8 0 0		

10. 467 + 33 = _____

7		3
6 0	+	3 0
4 0 0		

11. 582 + 78 = _____

+

12. 639 + 83 = _____

+

13. 758 + 69 = _____

+

14. 846 + 77 = _____

+

15. 837 + 86 = _____

+

- **Answer these questions in the same way.**

Now try this!

834 + 62 = _____   625 + 57 = _____   156 + 83 = _____

736 + 58 = _____   236 + 82 = _____   873 + 64 = _____

636 + 73 = _____   867 + 37 = _____   516 + 85 = _____

Teachers' note Before beginning, discuss how partitioning numbers into hundreds, tens and units can be used to help find totals mentally.

**Developing Numeracy
Mental Maths Year 5**
© A & C BLACK

Picking pairs

- **Ring five pairs of** adjacent **numbers on each number line.**
- **Find the total of each pair. Use doubling to help you.**

> **Adjacent** means next to each other.

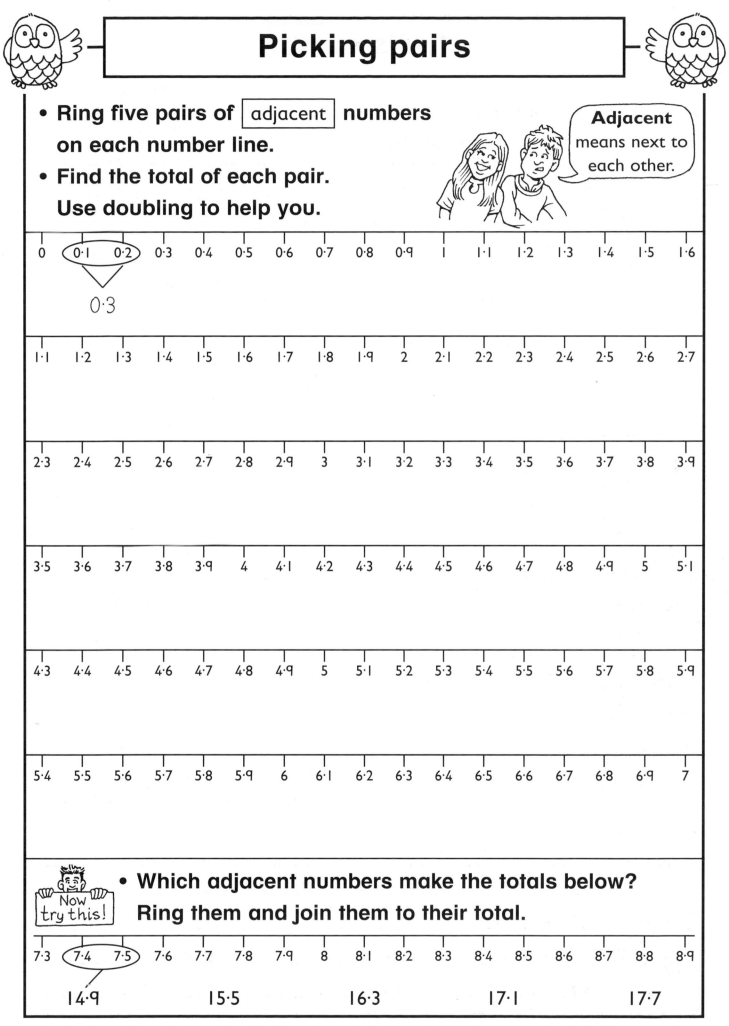

| 0 | 0·1 | 0·2 | 0·3 | 0·4 | 0·5 | 0·6 | 0·7 | 0·8 | 0·9 | 1 | 1·1 | 1·2 | 1·3 | 1·4 | 1·5 | 1·6 |

0·3

| 1·1 | 1·2 | 1·3 | 1·4 | 1·5 | 1·6 | 1·7 | 1·8 | 1·9 | 2 | 2·1 | 2·2 | 2·3 | 2·4 | 2·5 | 2·6 | 2·7 |

| 2·3 | 2·4 | 2·5 | 2·6 | 2·7 | 2·8 | 2·9 | 3 | 3·1 | 3·2 | 3·3 | 3·4 | 3·5 | 3·6 | 3·7 | 3·8 | 3·9 |

| 3·5 | 3·6 | 3·7 | 3·8 | 3·9 | 4 | 4·1 | 4·2 | 4·3 | 4·4 | 4·5 | 4·6 | 4·7 | 4·8 | 4·9 | 5 | 5·1 |

| 4·3 | 4·4 | 4·5 | 4·6 | 4·7 | 4·8 | 4·9 | 5 | 5·1 | 5·2 | 5·3 | 5·4 | 5·5 | 5·6 | 5·7 | 5·8 | 5·9 |

| 5·4 | 5·5 | 5·6 | 5·7 | 5·8 | 5·9 | 6 | 6·1 | 6·2 | 6·3 | 6·4 | 6·5 | 6·6 | 6·7 | 6·8 | 6·9 | 7 |

Now try this!

- **Which adjacent numbers make the totals below? Ring them and join them to their total.**

| 7·3 | 7·4 | 7·5 | 7·6 | 7·7 | 7·8 | 7·9 | 8 | 8·1 | 8·2 | 8·3 | 8·4 | 8·5 | 8·6 | 8·7 | 8·8 | 8·9 |

14·9   15·5   16·3   17·1   17·7

Teachers' note Discuss how knowledge of doubles can be used to find the totals. Invite the children to explain their strategies to a partner: for example, 'I know that double 27 is 54, so double 2·7 must be 5·4. This tells me that the answer to 2·7 + 2·8 must be 0·1 more than 5·4, which is 5·5.'

**Developing Numeracy
Mental Maths Year 5
© A & C BLACK**

19

Wrap up well

• **Choose your own three-digit number to write at the top of each scarf. Then fill in the missing numbers.**

1.
524
+ 29
553
− 41
− 69
+ 81

2.
+ 39
+ 29
− 41

3.
+ 21
− 49
+ 52
− 38

4.
+ 51
− 29
− 58

5.
− 69
+ 41
− 72
− 58

6.
+ 38
− 59
− 69
− 42

7.
− 68
+ 79
− 41
− 78

8.
+ 28
− 39
− 49
− 51
− 58
+ 79

Now try this!

• **For each scarf, find the** difference **between the number at the top and the number at the bottom.**

Teachers' note Demonstrate the strategy of adding or subtracting the nearest multiple of 10 and adjusting: for example, 524 + 29 can be found by adding 30 and subtracting 1. The answers to the extension activity can be used as a quick way of marking the children's work, whatever three-digit numbers they have chosen.

Developing Numeracy
Mental Maths Year 5
© A & C BLACK

Animal anagrams

- Find the total of the numbers on each card.
- Cut out the cards. Sort them into piles with the same total. Then order the letters in each pile to spell the name of a wild animal.

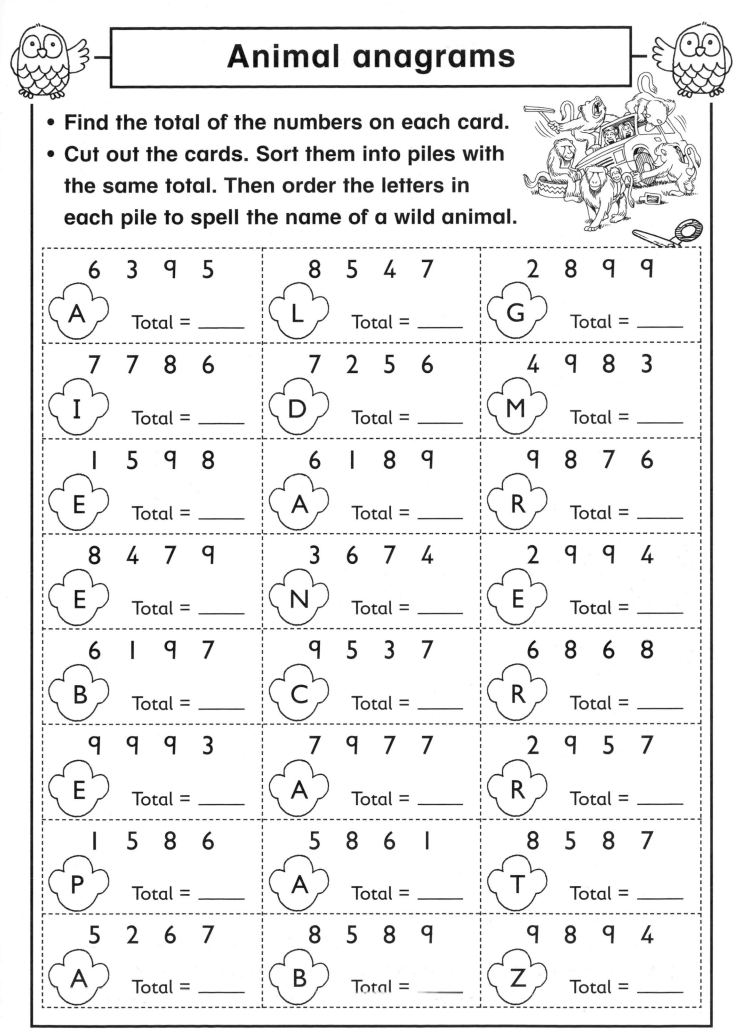

6 3 9 5 **A** Total = ___	8 5 4 7 **L** Total = ___	2 8 9 9 **G** Total = ___
7 7 8 6 **I** Total = ___	7 2 5 6 **D** Total = ___	4 9 8 3 **M** Total = ___
1 5 9 8 **E** Total = ___	6 1 8 9 **A** Total = ___	9 8 7 6 **R** Total = ___
8 4 7 9 **E** Total = ___	3 6 7 4 **N** Total = ___	2 9 9 4 **E** Total = ___
6 1 9 7 **B** Total = ___	9 5 3 7 **C** Total = ___	6 8 6 8 **R** Total = ___
9 9 9 3 **E** Total = ___	7 9 7 7 **A** Total = ___	2 9 5 7 **R** Total = ___
1 5 8 6 **P** Total = ___	5 8 6 1 **A** Total = ___	8 5 8 7 **T** Total = ___
5 2 6 7 **A** Total = ___	8 5 8 9 **B** Total = ___	9 8 9 4 **Z** Total = ___

Teachers' note Encourage checking by adding in a different order, and invite the children to say which order of adding they found easiest each time. As an extension activity, the children could create their own set of animal cards by finding different sets of numbers with the same total and writing a letter on each card to spell the name of an animal.

**Developing Numeracy
Mental Maths Year 5
© A & C BLACK**

Cross out

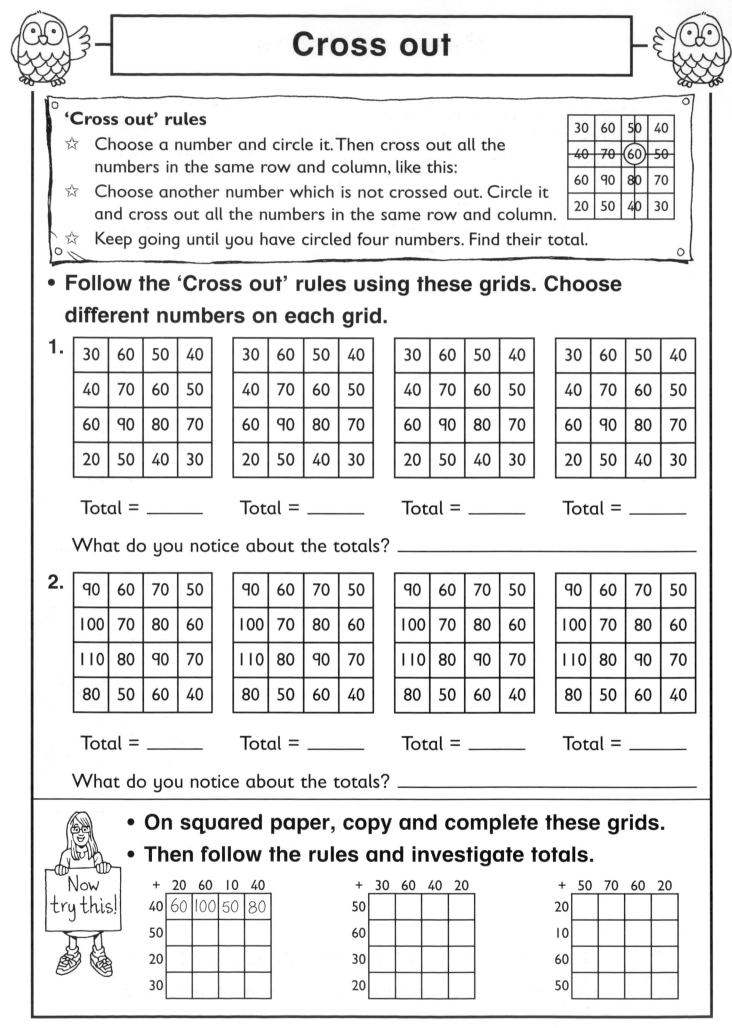

• **Follow the 'Cross out' rules using these grids. Choose different numbers on each grid.**

1.

30	60	50	40
40	70	60	50
60	90	80	70
20	50	40	30

30	60	50	40
40	70	60	50
60	90	80	70
20	50	40	30

30	60	50	40
40	70	60	50
60	90	80	70
20	50	40	30

30	60	50	40
40	70	60	50
60	90	80	70
20	50	40	30

Total = _____   Total = _____   Total = _____   Total = _____

What do you notice about the totals? _____

2.

90	60	70	50
100	70	80	60
110	80	90	70
80	50	60	40

90	60	70	50
100	70	80	60
110	80	90	70
80	50	60	40

90	60	70	50
100	70	80	60
110	80	90	70
80	50	60	40

90	60	70	50
100	70	80	60
110	80	90	70
80	50	60	40

Total = _____   Total = _____   Total = _____   Total = _____

What do you notice about the totals? _____

• **On squared paper, copy and complete these grids.**
• **Then follow the rules and investigate totals.**

Now try this!

+	20	60	10	40
40	60	100	50	80
50				
20				
30				

+	30	60	40	20
50				
60				
30				
20				

+	50	70	60	20
20				
10				
60				
50				

Teachers' note Ensure the children understand the 'Cross out' rules. They will need squared paper for the extension activity. Check that they understand how to fill in the grids and explain that they should copy out each grid several times to investigate the totals. In each grid the total of the circled numbers will be the sum of all the numbers around the edge, no matter which numbers are circled.

Developing Numeracy Mental Maths Year 5 © A & C BLACK

Petal patterns

☆ Write the numbers **21** to **29** in the squares of the grid, in any order. **Example:**

☆ Add pairs of **adjacent** numbers and write the sum in the petals. Then find the total of all the petal numbers.

☆ Try putting the numbers 21 to 29 in a different order. What is the highest total you can make?

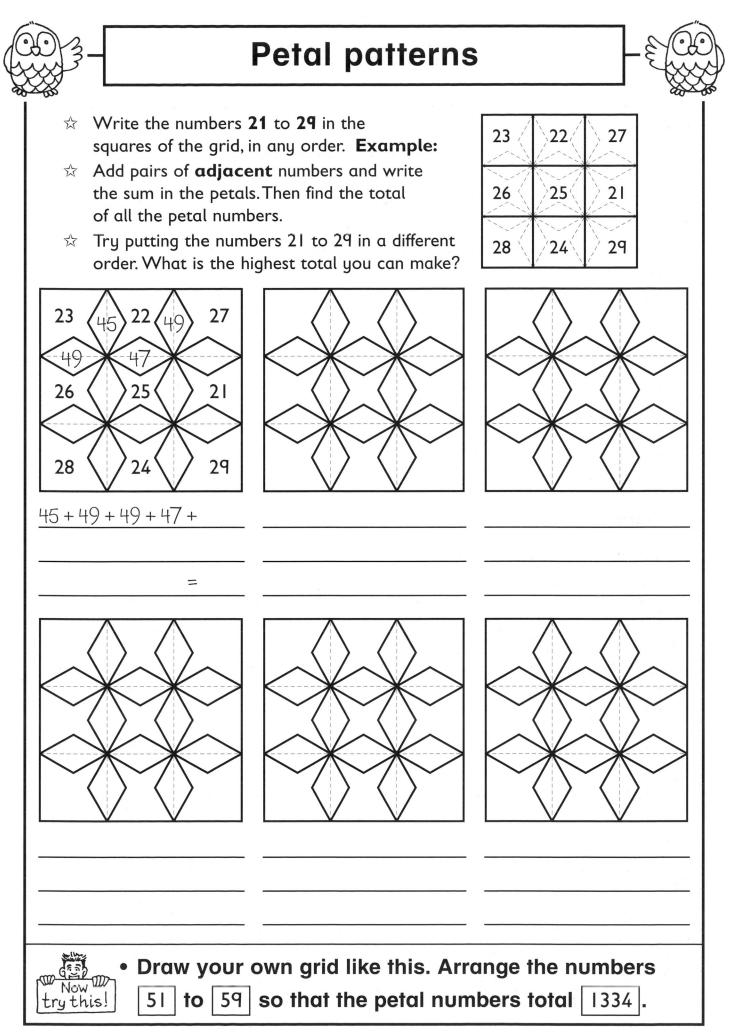

23	22	27
26	25	21
28	24	29

45 + 49 + 49 + 47 +

_____

=

• **Draw your own grid like this. Arrange the numbers** 51 **to** 59 **so that the petal numbers total** 1334 .

Now try this!

Teachers' note Remind the children of the meaning of 'adjacent' and discuss how the petal numbers can be added by making 'lots of 50' and then adjusting. Encourage them to be systematic and to think of reasons why some arrangements make higher totals. Provide squared paper for the extension activity. The children could also investigate the numbers 31 to 39, 61 to 69 or 71 to 79.

Developing Numeracy
Mental Maths Year 5
© A & C BLACK

Swap the digits

Each digit can be swapped for another digit, using this table.

0	1	2	3	4	5	6	7	8	9
↓	↓	↓	↓	↓	↓	↓	↓	↓	↓
8	9	0	1	2	3	4	5	6	7

The number 13 becomes 91, and the number 25 becomes 03 or 3.

- **Rewrite these questions by swapping the digits.**
- **Answer the original question <u>and</u> the question with swapped digits. Find the difference between the two answers.**

1. 5 7 + 3 6 = _93_

35 + 14 = 49

Diff 44

2. 4 3 + 5 2 = _____

____ + ____ = _____

Diff

3. 3 8 + 2 3 = _____

____ + ____ = _____

Diff

4. 3 7 + 6 2 = _____

____ + ____ = _____

Diff

5. 4 9 + 3 8 = _____

____ + ____ = _____

Diff

6. 7 4 + 2 6 = _____

____ + ____ = _____

Diff

7. 6 3 + 7 8 = _____

____ + ____ = _____

Diff

8. 2 9 + 9 5 = _____

____ + ____ = _____

Diff

- **What do you notice about the differences?** _____

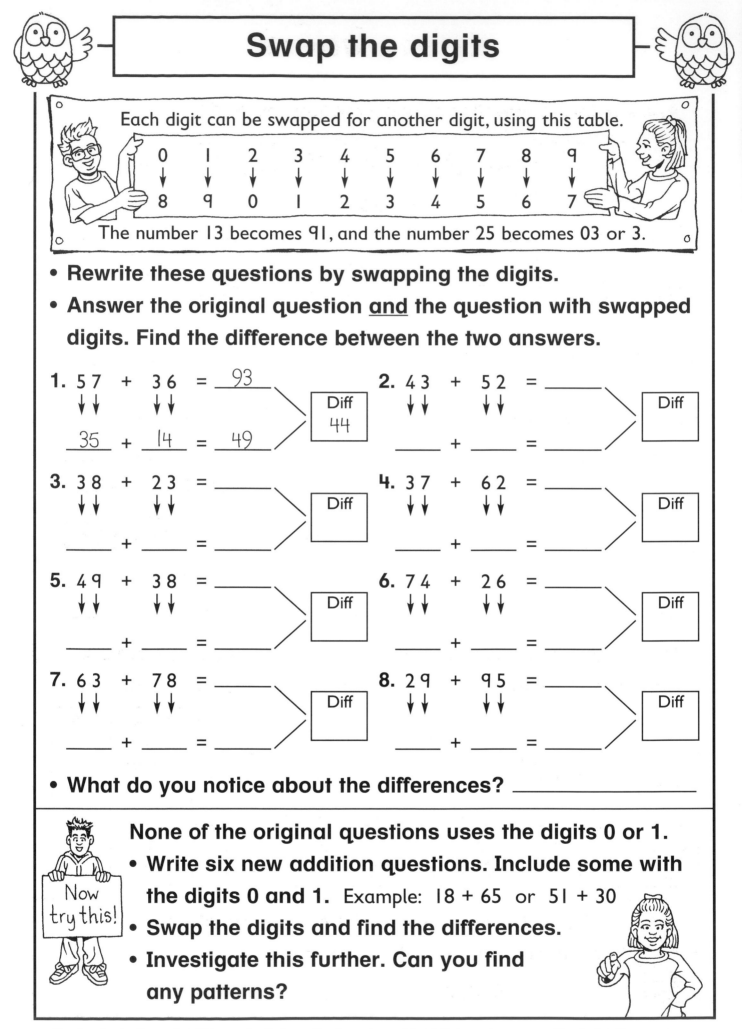

None of the original questions uses the digits 0 or 1.
- **Write six new addition questions. Include some with the digits 0 and 1.** Example: 18 + 65 or 51 + 30
- **Swap the digits and find the differences.**
- **Investigate this further. Can you find any patterns?**

Teachers' note Ensure the children realise that once they have swapped the digits to create a new question, they do not do any more swapping. In the extension activity, encourage them to group their questions according to the differences and to describe the patterns. Questions involving a 0 or 1 digit often produce the difference 34; questions with two 0 or 1 digits tend to produce a difference of 24.

**Developing Numeracy
Mental Maths Year 5
© A & C BLACK**

Ocean waves

- **Answer these questions using mental methods.**

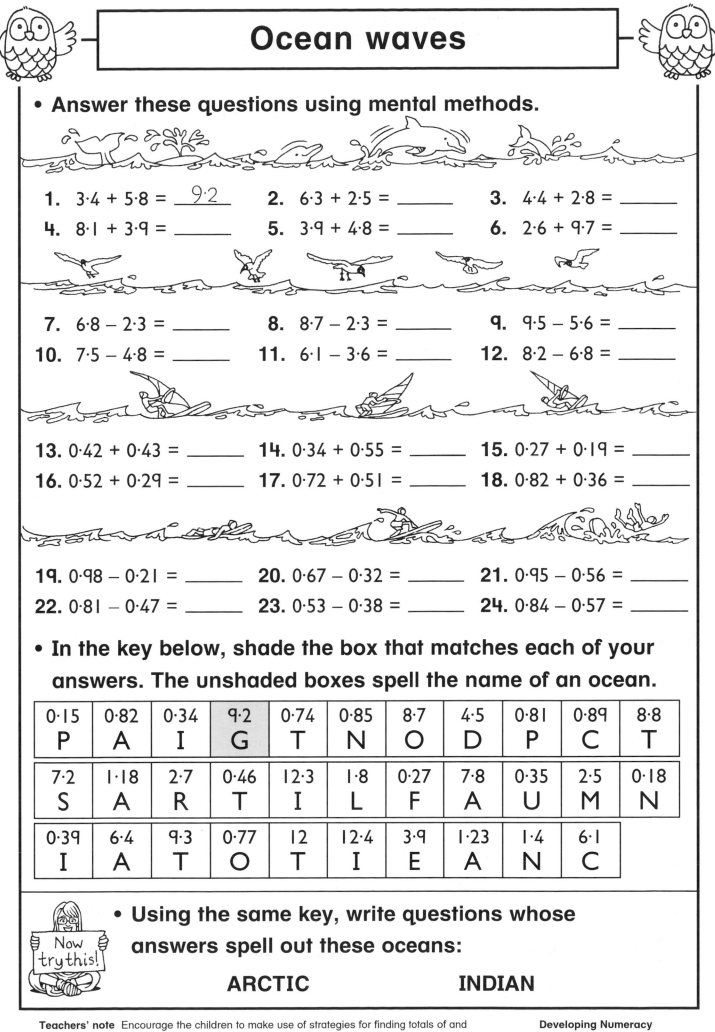

1. $3.4 + 5.8 =$ _9.2_   2. $6.3 + 2.5 =$ _____   3. $4.4 + 2.8 =$ _____
4. $8.1 + 3.9 =$ _____   5. $3.9 + 4.8 =$ _____   6. $2.6 + 9.7 =$ _____

7. $6.8 - 2.3 =$ _____   8. $8.7 - 2.3 =$ _____   9. $9.5 - 5.6 =$ _____
10. $7.5 - 4.8 =$ _____   11. $6.1 - 3.6 =$ _____   12. $8.2 - 6.8 =$ _____

13. $0.42 + 0.43 =$ _____ 14. $0.34 + 0.55 =$ _____ 15. $0.27 + 0.19 =$ _____
16. $0.52 + 0.29 =$ _____ 17. $0.72 + 0.51 =$ _____ 18. $0.82 + 0.36 =$ _____

19. $0.98 - 0.21 =$ _____ 20. $0.67 - 0.32 =$ _____ 21. $0.95 - 0.56 =$ _____
22. $0.81 - 0.47 =$ _____ 23. $0.53 - 0.38 =$ _____ 24. $0.84 - 0.57 =$ _____

- **In the key below, shade the box that matches each of your answers. The unshaded boxes spell the name of an ocean.**

0.15	0.82	0.34	9.2	0.74	0.85	8.7	4.5	0.81	0.89	8.8
P	A	I	G	T	N	O	D	P	C	T

7.2	1.18	2.7	0.46	12.3	1.8	0.27	7.8	0.35	2.5	0.18
S	A	R	T	I	L	F	A	U	M	N

0.39	6.4	9.3	0.77	12	12.4	3.9	1.23	1.4	6.1
I	A	T	O	T	I	E	A	N	C

- **Using the same key, write questions whose answers spell out these oceans:**

ARCTIC   **INDIAN**

Teachers' note Encourage the children to make use of strategies for finding totals of and differences between pairs of two-digit whole numbers: for example, the same approach can be used to answer 2·6 + 4·9 as for 26 + 49. Suggest that the children check their answers using a different calculation, such as subtraction to check addition, and vice versa.

Developing Numeracy Mental Maths Year 5 © A & C BLACK

Multiply to add

• **Use multiplication to help you add these numbers.**

1. | 10 | 11 | 10 | 11 | 9 |

$5 \times 10 = 50$

+1   +1   −1

Total = __51__

*Each of these five numbers is about **10**. So, I can multiply by 10 and then adjust.*

2. | 20 | 21 | 19 | 19 |

$4 \times 20 =$

Total = _____

3. | 12 | 11 | 10 | 11 | 12 |

Total = _____

4. | 31 | 31 | 29 | 30 | 28 |

Total = _____

5. | 40 | 42 | 40 | 43 | 39 |

Total = _____

6. | 15 | 16 | 13 | 14 |

Total = _____

7. | 21 | 20 | 22 | 18 | 19 |

Total = _____

8. | 49 | 50 | 48 | 52 | 50 |

Total = _____

9. | 50 | 48 | 51 | 52 | 49 | 50 |

Total = _____

10. | 27 | 26 | 28 | 27 |

Total = _____

11. | 40 | 41 | 44 | 40 | 38 | 39 |

Total = _____

• **Find the** perimeter **of each shape using a similar method.**

Now try this!

1·7 cm
1·5 cm
1·8 cm
1·4 cm   1·5 cm

5·2 cm
5 cm   4·9 cm
4·9 cm   5·1 cm
4·9 cm

• **Sketch and label two shapes for a partner to solve.**

Teachers' note Demonstrate how the totals can be found by multiplying and then adjusting. Discuss that '+1' and '−1' cancel each other out. Ensure the children realise that they could choose any number for the approximation. For the extension activity, they may need reminding that perimeter is the distance once around the edge of a shape.

**Developing Numeracy
Mental Maths Year 5
© A & C BLACK**

Sum and difference

- **Start with the two numbers on the flowers. Find the** sum **and the** difference **. Write the answers on the leaves.**
- **Now find the sum of these two answers and the difference between them. Keep going down the flower stem.**

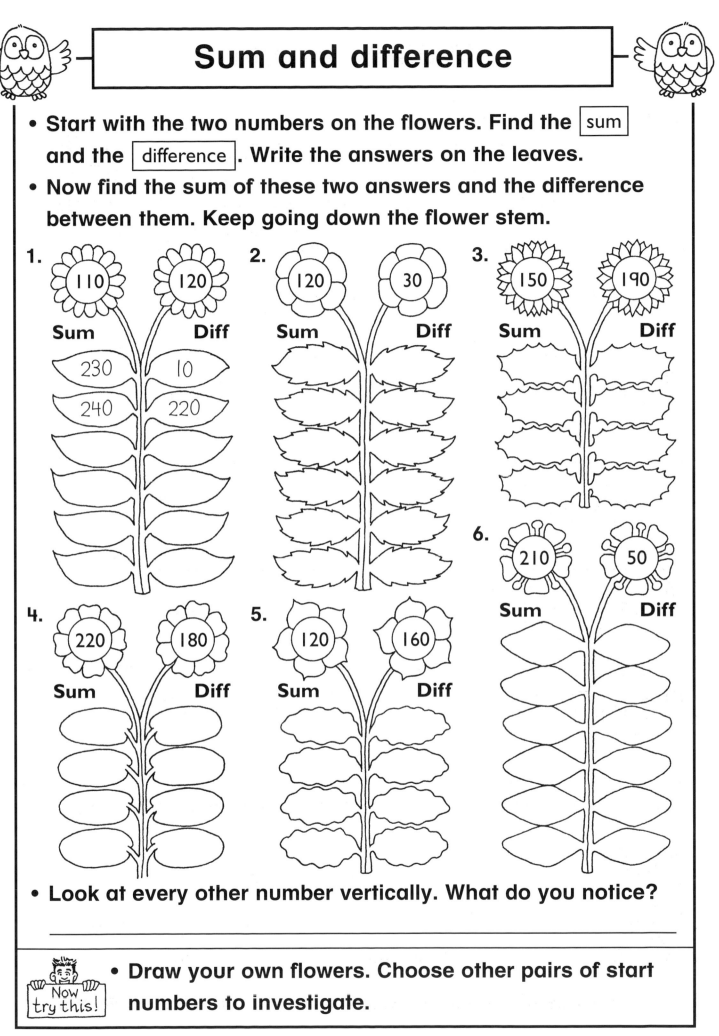

1. 110   120
   Sum   Diff
   230   10
   240   220

2. 120   30
   Sum   Diff

3. 150   190
   Sum   Diff

4. 220   180
   Sum   Diff

5. 120   160
   Sum   Diff

6. 210   50
   Sum   Diff

- **Look at every other number vertically. What do you notice?**

- **Draw your own flowers. Choose other pairs of start numbers to investigate.**

Teachers' note This sheet provides practice in adding and subtracting pairs of two- and three-digit multiples of 10. Ensure the children realise that they should find the sum of and difference between the two previous answers each time. The children could be asked to write the longest string of sums and differences that they can, using only mental methods.

Developing Numeracy Mental Maths Year 5 © A & C BLACK

27

Quick-fire questions

- **Time yourself as you answer each round of questions.**

Round 1

1. Multiply nine by two ☐
2. Four lots of seven ☐
3. Nine nines ☐
4. Five times nine ☐
5. Six groups of eight ☐
6. Three squared ☐
7. The product of seven and three ☐
8. Two multiplied by six ☐

Time _____

Round 2

1. Multiply six by six ☐
2. Six lots of seven ☐
3. Five fives ☐
4. Four times eight ☐
5. Four groups of nine ☐
6. Four squared ☐
7. The product of eight and five ☐
8. Five multiplied by six ☐

Time _____

Round 3

1. Multiply eight by six ☐
2. Seven lots of six ☐
3. Ten tens ☐
4. Six times four ☐
5. Nine groups of eight ☐
6. Seven squared ☐
7. The product of four and nine ☐
8. Five multiplied by seven ☐

Time _____

Round 4

1. Multiply seven by six ☐
2. Eight lots of seven ☐
3. Four fours ☐
4. Three times nine ☐
5. Six groups of five ☐
6. Eight squared ☐
7. The product of six and eight ☐
8. Seven multiplied by nine ☐

Time _____

- **In which round were you quickest?** _____

Teachers' note At the start of the lesson, revise that squaring a number means multiplying it by itself, and that to find the product you multiply numbers together. The children could use stopwatches or clocks to time themselves. As an extension activity, ask the children to make up their own round of questions, using the 6, 7 and 8 times tables, for a partner to answer.

**Developing Numeracy
Mental Maths Year 5
© A & C BLACK**

Juggle your tables

- **Multiply** adjacent **numbers on the balls. Find the total of your answers.**

Which balls have the highest and lowest totals?

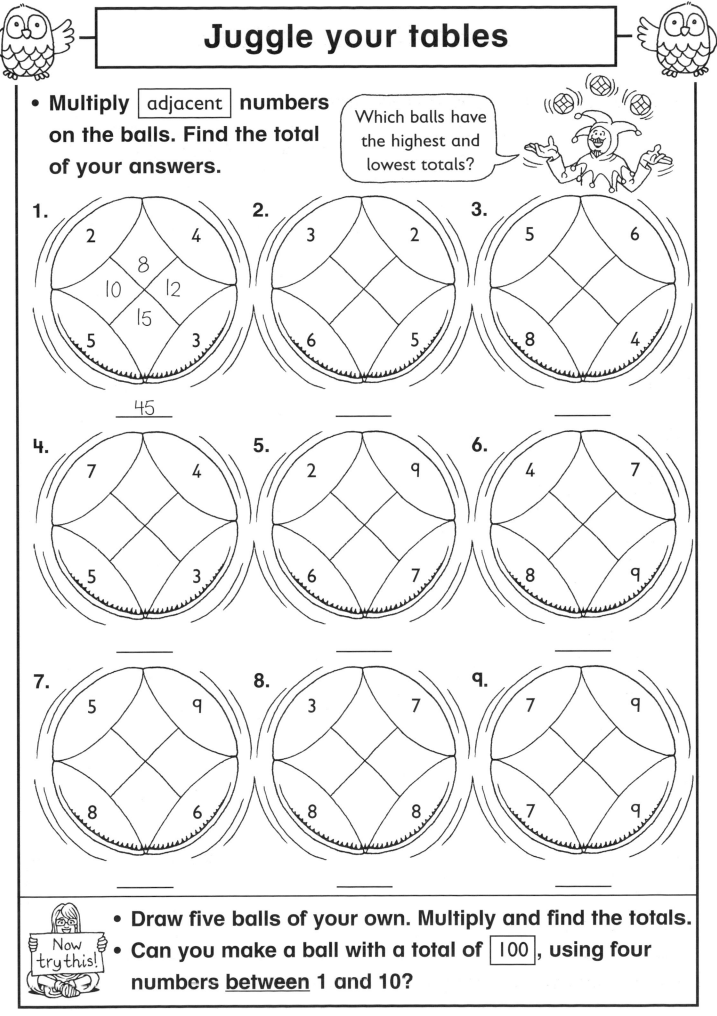

1. 2 4 8 10 12 15 5 3

_____ 45

2. 3 2 6 5

_____

3. 5 6 8 4

_____

4. 7 4 5 3

_____

5. 2 9 6 7

_____

6. 4 7 8 9

_____

7. 5 9 8 6

_____

8. 3 7 8 8

_____

9. 7 9 7 9

_____

- **Draw five balls of your own. Multiply and find the totals.**
- **Can you make a ball with a total of** 100 **, using four numbers <u>between</u> 1 and 10?**

Now try this!

Teachers' note Demonstrate how to complete the balls, explaining the term 'adjacent'. Discuss methods for finding the totals, such as grouping numbers that make 10 or 20, or adding the tens before counting on the units. The children can also investigate whether it matters where numbers are placed on the ball. Ask: 'What is the lowest/highest total you can make using 2, 3, 4 and 5?'

**Developing Numeracy
Mental Maths Year 5**
© A & C BLACK

Consecutive patterns

- **Choose three** consecutive **numbers less than 15. Write them on the shamrock in order, like this:**

- **Multiply the top two numbers. Then square the bottom number.**

- **Write in words what you notice.**

_____

_____

Will this work for all consecutive numbers?

- **Investigate three consecutive even numbers or three consecutive multiples of 10.**

  Example: 4 6 8 **or** 30 40 50

- **Describe the patterns you notice.**

Teachers' note This activity provides practice in multiplying and squaring and can be widened to include any number range. Ensure that the children understand the term 'consecutive' and check that they write the numbers on the shamrocks in the correct order. Encourage them to describe clearly the patterns they notice and to suggest reasons for them.

Developing Numeracy
Mental Maths Year 5
© A & C BLACK

Apple trail

• **Play this game with a partner.**

☆ **You need** a dice and two small counters.

☆ Take turns to roll the dice and move your counter. Work out the division and find the answer further along the trail. Move your counter to this apple and wait until your next turn to roll the dice again.

☆ The first player to munch through all the apples is the winner.

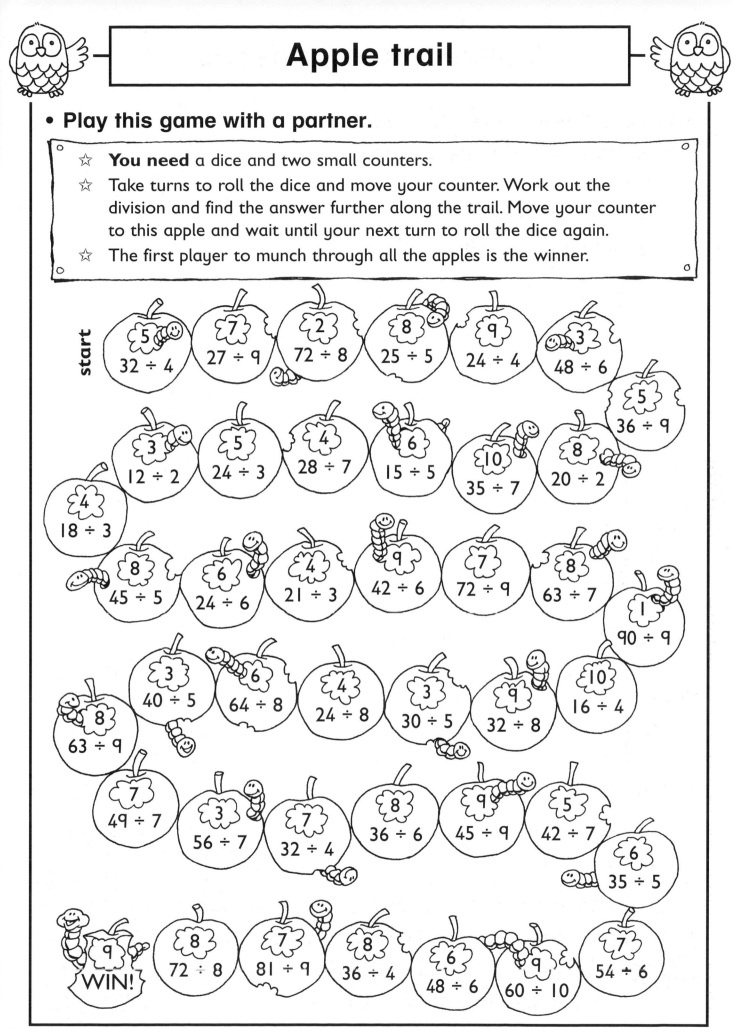

Teachers' note Some children may require a list of times tables facts to help them derive the answers.

Developing Numeracy
Mental Maths Year 5
© A & C BLACK

On the double!

This machine doubles numbers.

First it splits the number.

Then it doubles each part.

Finally, it adds them together.

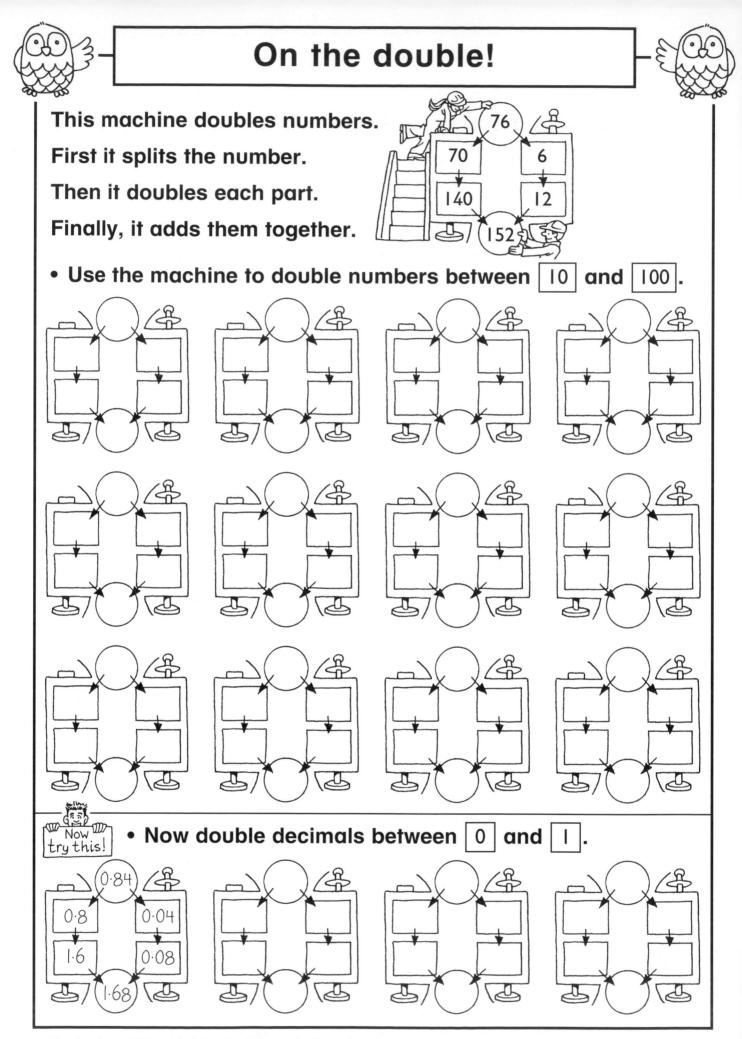

* Use the machine to double numbers between ⌐10⌐ and ⌐100⌐.

* Now double decimals between ⌐0⌐ and ⌐1⌐.

Teachers' note This activity helps the children to double numbers to 100, and to memorise them. Discuss that the machine splits the number into tens and units. The numbers could be filled in before photocopying (including numbers where the tens and/or units digit is 5, 6, 7, 8 or 9). Alternatively, the children can choose numbers, or generate them by picking digit cards.

**Developing Numeracy
Mental Maths Year 5
© A & C BLACK**

Double take!

In a game show, the number of points scored is doubled and a cheque is written for that amount.

• **Double these points and complete the cheques.**

Double 400
Pay IVOR LOTT £ 800
123456789-1

Double 700
Pay TERESA GREEN £
123456789-2

Double 1100
Pay ROBIN FEATHERS £
123456789-3

Double 1300
Pay TONY BROKE £
123456789-4

Double 1600
Pay LILY SCUTTLE £
123456789-5

Double 1900
Pay BRIAN BUTTERFINGERS £
123456789-6

Double 2400
Pay VINCE VAN GOGH £
123456789-7

WINNER!

Double 3500
Pay PENNY FARTHING £
123456789-8

Double 4700
Pay SALLY DRIBBLES £
123456789-9

Double 5100
Pay MAJOR CHEAT £
123456789-10

Double 5600
Pay TOBY TENPENCE £
123456789-11

Double 6700
Pay HILDA HARDUP £
123456789-12

Double 7300
Pay BERT SCRAGGINS £
123456789-13

Double 8500
Pay FRANK N. STEIN £
123456789-14

Double 8900
Pay PROF. PLUM £
123456789-15

Double 9600
Pay JONNY CASH £
123456789-16

Now try this!

• **Halve these amounts to find the points scored.**

£13 800 ↓ ____   £17 600 ↓ ____   £19 000 ↓ ____   £18 800 ↓ ____   £19 400 ↓ ____

Teachers' note This activity could be timed to encourage rapid recall of facts. Encourage the children to derive facts they do not know by using known facts. Ask them to describe their strategies to a partner: for example, 'I know that double 51 is 102, so double 5100 must be 10 200.'

**Developing Numeracy
Mental Maths Year 5
© A & C BLACK**

33

The long and the short of it

- **Halve** the start number. Keep halving your answer until you reach a number that ends in 5. One has been done for you.

Which start numbers make the longest and shortest chains?

1440	1040	1920	1760	1280	1360	19 200	16 000	12 800
720								
360								
180								
90								
45								

Now try this!

- **Keep** doubling **each of these finish numbers. Then join the finish number to its start number.**

Finish: 575   115   225   525   1125   105

Start: 1680   16 800   1840   18 400   14 400   144 000

Teachers' note This activity provides practice in halving multiples of 10 and 100 up to 20 000. Encourage the children to describe the strategies they use when halving, such as splitting the number into parts and halving each part before recombining.

Developing Numeracy Mental Maths Year 5 © A & C BLACK

Thinking caps on!

- To answer these multiplications, ⎡double⎤ one number and ⎡halve⎤ the other. Think carefully about which to double and which to halve.

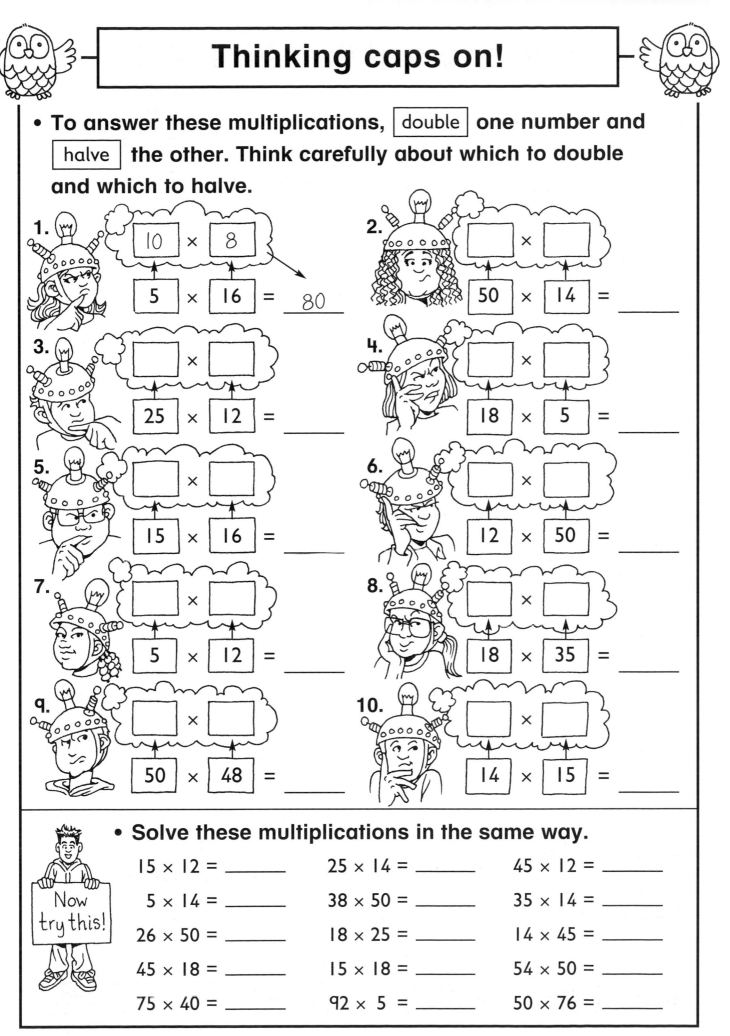

1.
$$\boxed{10} \times \boxed{8}$$
$$\boxed{5} \times \boxed{16} = \underline{80}$$

2.
$$\boxed{} \times \boxed{}$$
$$\boxed{50} \times \boxed{14} = \underline{}$$

3.
$$\boxed{} \times \boxed{}$$
$$\boxed{25} \times \boxed{12} = \underline{}$$

4.
$$\boxed{} \times \boxed{}$$
$$\boxed{18} \times \boxed{5} = \underline{}$$

5.
$$\boxed{} \times \boxed{}$$
$$\boxed{15} \times \boxed{16} = \underline{}$$

6.
$$\boxed{} \times \boxed{}$$
$$\boxed{12} \times \boxed{50} = \underline{}$$

7.
$$\boxed{} \times \boxed{}$$
$$\boxed{5} \times \boxed{12} = \underline{}$$

8.
$$\boxed{} \times \boxed{}$$
$$\boxed{18} \times \boxed{35} = \underline{}$$

9.
$$\boxed{} \times \boxed{}$$
$$\boxed{50} \times \boxed{48} = \underline{}$$

10.
$$\boxed{} \times \boxed{}$$
$$\boxed{14} \times \boxed{15} = \underline{}$$

- Solve these multiplications in the same way.

Now try this!

$15 \times 12 = \underline{}$ $25 \times 14 = \underline{}$ $45 \times 12 = \underline{}$

$5 \times 14 = \underline{}$ $38 \times 50 = \underline{}$ $35 \times 14 = \underline{}$

$26 \times 50 = \underline{}$ $18 \times 25 = \underline{}$ $14 \times 45 = \underline{}$

$45 \times 18 = \underline{}$ $15 \times 18 = \underline{}$ $54 \times 50 = \underline{}$

$75 \times 40 = \underline{}$ $92 \times 5 = \underline{}$ $50 \times 76 = \underline{}$

Teachers' note At the start of the lesson, revise multiplying multiples of 10 using tables facts as a starting point (for example, 9 × 30, 6 × 20, 30 × 50). Demonstrate how to double one number and halve the other, and show that this produces the same answer as the original question.

**Developing Numeracy
Mental Maths Year 5
© A & C BLACK**

Cut and sort

- **Cut out the cards and the three labels.**
- **Sort the cards using the labels.**
- **Swap places with a partner and check each other's answers.**

Cards with the answer **24**	Cards with the answer **32**	Cards with the answer **36**
$2 \times 6 \times 2$	$2 \times 8 \times 2$	$4 \times 3 \times 2$
$3 \times 2 \times 2 \times 2$	$3 \times 3 \times 2 \times 2$	6×6
6×4	2×18	$6 \times 2 \times 3$
$2 \times 4 \times 4$	$3 \times 4 \times 3$	16×2
$4 \times 2 \times 2 \times 2$	4×8	4×6
2×16	12×2	$48 \div 2$
$72 \div 2$	$64 \div 2$	$72 \div 3$
$96 \div 3$	$(72 \div 4) \times 2$	$(64 \div 8) \times 4$
$4 + (100 \div 5)$	$6 + (5 \times 6)$	$12 + (4 \times 5)$
$40 - (2 \times 8)$	$40 - (4 \times 4)$	$40 - (36 \div 9)$
$(50 \div 2) - 1$	$5 + (38 \div 2)$	$4 + (7 \times 4)$
$(17 \times 2) - 2$	$(17 \times 2) + 2$	$(81 \div 9) \times 4$

Teachers' note Revise the use of brackets and discuss that three part inside the brackets must be done first: for example, $6 + (5 \times 6)$ is the same as $6 + 30$. The children should be encouraged to discuss the strategies they use to find the answers. As an extension, ask the children to write five more statements for each pile, using brackets.

**Developing Numeracy
Mental Maths Year 5
© A & C BLACK**

Windy moments

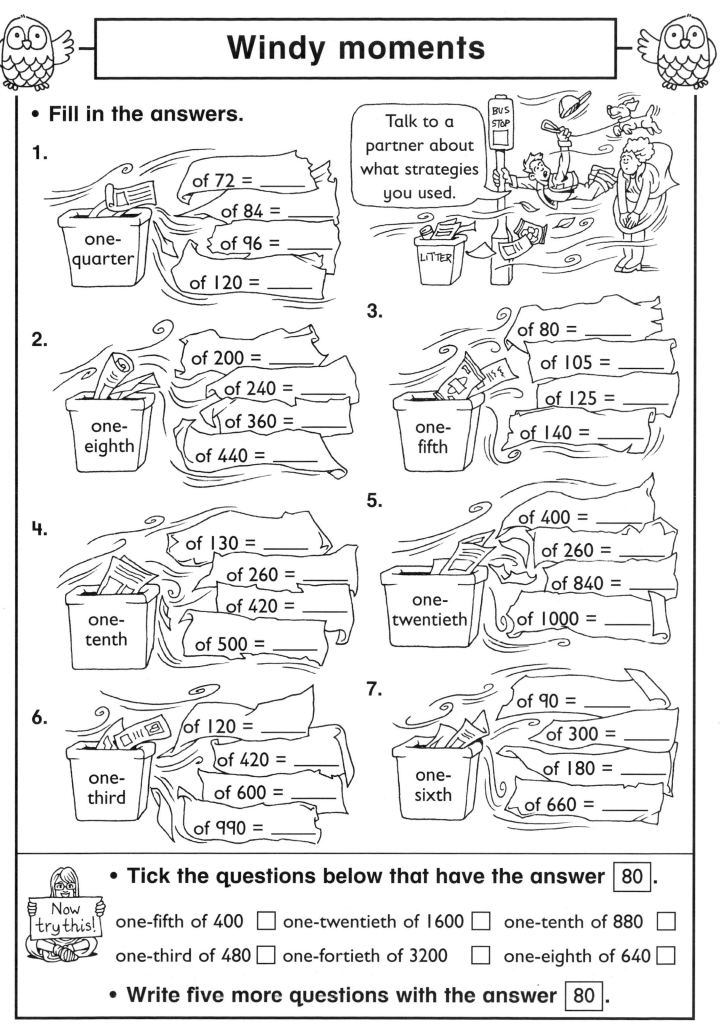

• **Fill in the answers.**

Talk to a partner about what strategies you used.

1. one-quarter

of 72 = _____
of 84 = _____
of 96 = _____
of 120 = _____

2. one-eighth

of 200 = _____
of 240 = _____
of 360 = _____
of 440 = _____

3. one-fifth

of 80 = _____
of 105 = _____
of 125 = _____
of 140 = _____

4. one-tenth

of 130 = _____
of 260 = _____
of 420 = _____
of 500 = _____

5. one-twentieth

of 400 = _____
of 260 = _____
of 840 = _____
of 1000 = _____

6. one-third

of 120 = _____
of 420 = _____
of 600 = _____
of 990 = _____

7. one-sixth

of 90 = _____
of 300 = _____
of 180 = _____
of 660 = _____

Now try this!

• **Tick the questions below that have the answer** 80 .

one-fifth of 400 ☐   one-twentieth of 1600 ☐   one-tenth of 880 ☐

one-third of 480 ☐   one-fortieth of 3200 ☐   one-eighth of 640 ☐

• **Write five more questions with the answer** 80 .

Teachers' note This activity requires the children to solve a range of mental divisions, and to make decisions about how best to solve them. Possible strategies include using halving and doubling (finding one-sixth of 300 is the same as finding one-third of 300 and then halving) and using knowledge of tables facts (for example, 6 × 7 = 42, so one-sixth of 420 is 70).

Developing Numeracy Mental Maths Year 5 © A & C BLACK

Clever cards

Each of the numbers below can be made using numbers and signs from the cards.

You can use the same card more than once.

2 = 5

7 × 3

- Write a multiplication for each number.

$6 =$ ___ 2×3 ___

$12 =$ ___ $2 \times 2 \times 3$ ___ $14 =$ _____

$15 =$ _____ $18 =$ _____

$21 =$ _____ $25 =$ _____

$42 =$ _____ $45 =$ _____

- Now use your multiplications above to help you answer these.

Remember, you can multiply in any order.

$6 \times 15 =$ ___ $2 \times 3 \times 3 \times 5 = 10 \times 9 = 90$ ___

$15 \times 14 =$ _____

$18 \times 15 =$ _____

$15 \times 12 =$ _____

$6 \times 25 =$ _____

$15 \times 42 =$ _____

$15 \times 21 =$ _____

$45 \times 12 =$ _____

$18 \times 45 =$ _____

$21 \times 25 =$ _____

Now try this!

- Look at this fact: $6 \times 15 = 90$
- Use it to help you write at least ten other multiplication or division facts.   Example: $7 \times 15 = 105$ $90 \div 15 = 6$

Teachers' note This activity explores the use of factors to make mental multiplication easier. At the start of the lesson, discuss that multiplication can be done in any order: for example, $2 \times 3 \times 5$ can be answered by multiplying the 2 by the 5 first and then multiplying by 3. The children could work in pairs for this activity.

Developing Numeracy Mental Maths Year 5 © A & C BLACK

First in the middle

- **Multiply the number on the trophy by** 19 , 20 **and** 21 .
  Multiply by 20 first, then use this to help you work out
  the other answers.

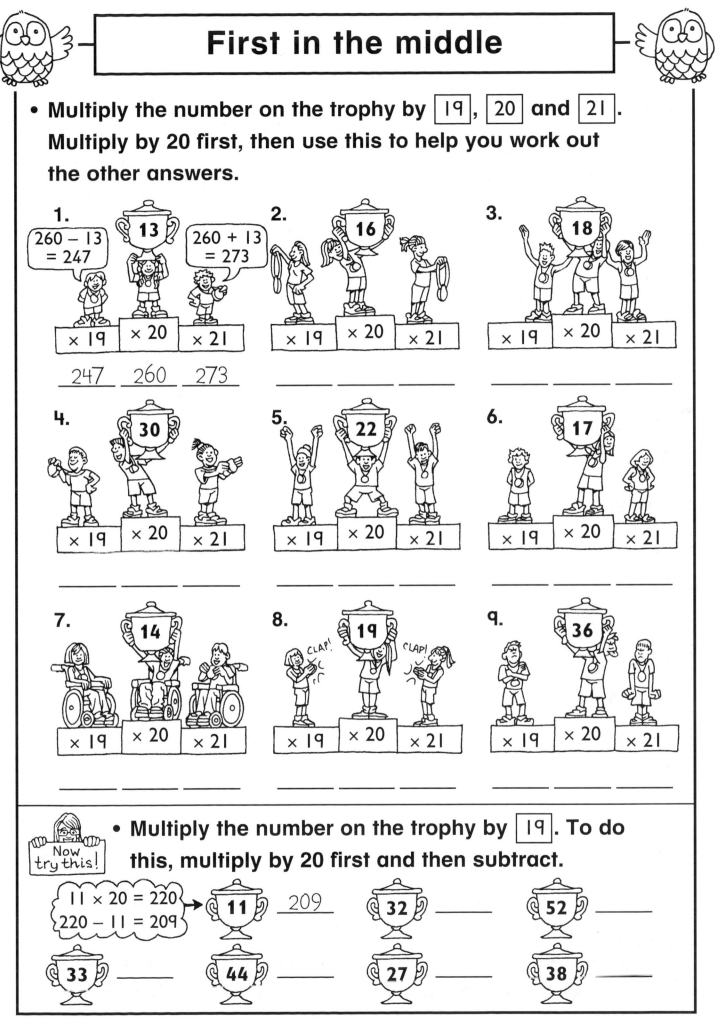

1.
260 − 13 = 247
13
260 + 13 = 273
× 19   × 20   × 21
247   260   273

2.
16
× 19   × 20   × 21

3.
18
× 19   × 20   × 21

4.
30
× 19   × 20   × 21

5.
22
× 19   × 20   × 21

6.
17
× 19   × 20   × 21

7.
14
× 19   × 20   × 21

8.
19
× 19   × 20   × 21

9.
36
× 19   × 20   × 21

Now try this!

- **Multiply the number on the trophy by** 19 . **To do**
  this, multiply by 20 first and then subtract.

11 × 20 = 220
220 − 11 = 209
→ **11**   209   **32** ____   **52** ____

33 ____   **44** ____   **27** ____   **38** ____

Teachers' note Demonstrate this strategy at the start of the lesson. Show that to multiply a number by 19 we can multiply by 20 and then subtract the number. To multiply a number by 21 we can multiply by 20 and then add the number. The children can check their answers by making sure that the three answers in each set go up in steps of equal size (the size of the number on the trophy).

Developing Numeracy
Mental Maths Year 5
© A & C BLACK

Oranges and lemons

Oranges: 47, 36, 62, 54, 73
Lemons: 3, 4, 5

- **Choose an orange and a lemon and multiply the numbers together. Split the first number into tens and units to help you.**
- **Which calculation gives an answer closest to** 200 **?**

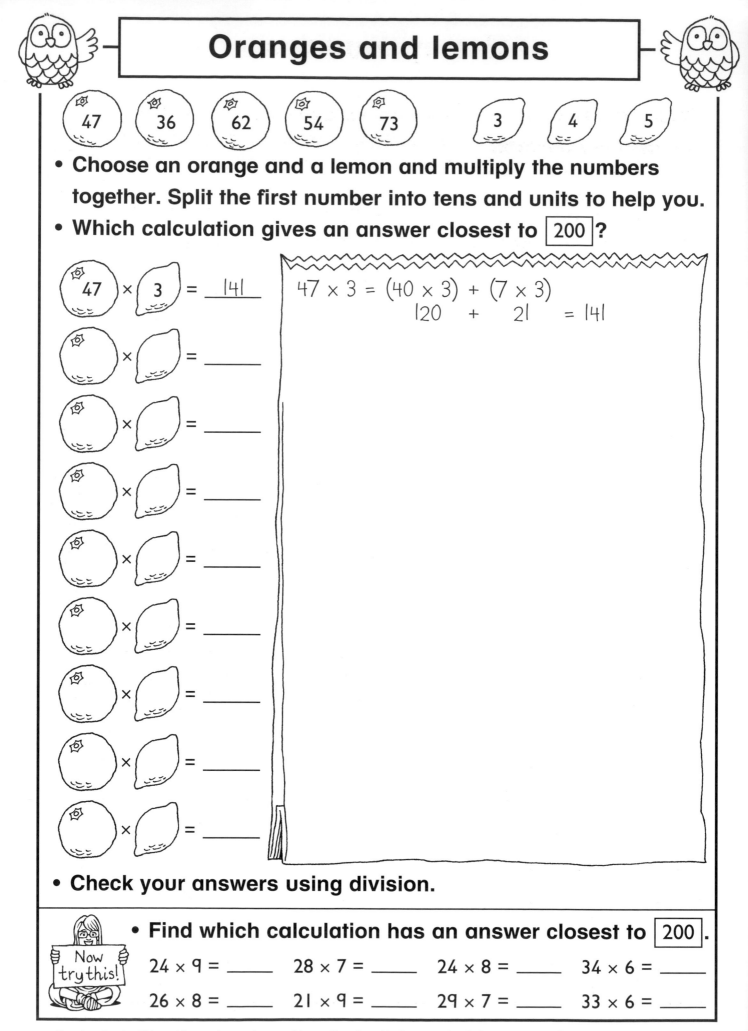

47 × 3 = __141__

47 × 3 = (40 × 3) + (7 × 3)
        120   +   21   = 141

( ) × ( ) = _____

( ) × ( ) = _____

( ) × ( ) = _____

( ) × ( ) = _____

( ) × ( ) = _____

( ) × ( ) = _____

( ) × ( ) = _____

( ) × ( ) = _____

- **Check your answers using division.**

- **Find which calculation has an answer closest to** 200 .

Now try this!

24 × 9 = _____   28 × 7 = _____   24 × 8 = _____   34 × 6 = _____

26 × 8 = _____   21 × 9 = _____   29 × 7 = _____   33 × 6 = _____

Teachers' note This activity can be used to provide practice of partitioning a number before multiplying and then combining the parts. At the start of the lesson, revise tables facts and multiplying a multiple of 10 by a single-digit number (such as 50 × 7). Show how partitioning can be used to reach answers more quickly. Encourage the children to make jottings.

Developing Numeracy Mental Maths Year 5 © A & C BLACK

Money madness

This is a game for two or three players.

You each need two counters of the same colour and some scrap paper to keep score.

☆ All players place one counter anywhere on the top grid and one anywhere on the bottom grid.

☆ Take turns to move your counters one square in any direction to create a division question. Work out the answer to the question and write it down. You only win money if the answer is a whole number of pounds. For an answer with a remainder you score 0.

☆ The winner is the first to reach £1000.

£100	£72	£54	£24	£70
£30	£10	£200	£36	£45
£63	£50	£78	£27	£150

÷ 100	÷ 4	÷ 3	÷ 5	÷ 9	÷ 10
÷ 9	÷ 8	÷ 4	÷ 7	÷ 8	÷ 7
÷ 3	÷ 100	÷ 50	÷ 3	÷ 100	÷ 2
÷ 5	÷ 7	÷ 4	÷ 8	÷ 9	÷ 50

Teachers' note Only one copy of the page is needed between two or three players. As an extension activity, the children could be asked to explore which squares produce the highest and lowest possible scores. Encourage the children to use jottings to help them, where necessary.

Developing Numeracy Mental Maths Year 5 © A & C BLACK

Follow the star

- **For each star, start at A with the number** $\boxed{10}$ **. Follow the arrows until you reach A again. Write the finish number.**

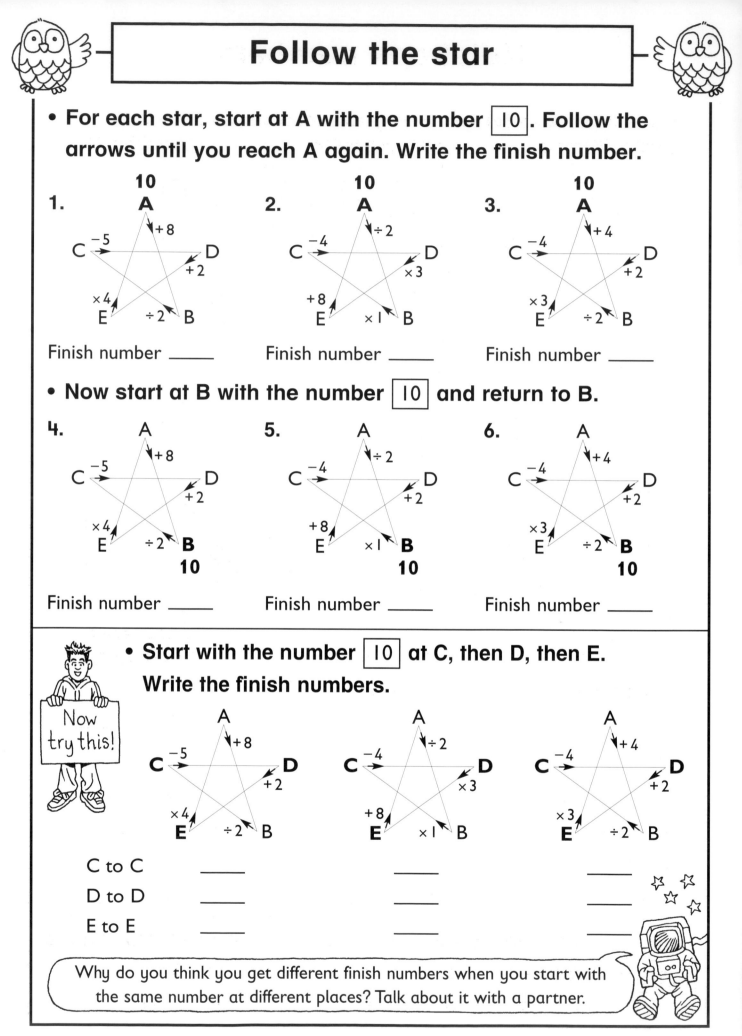

1.

10
A +8
C −5 → D
+2
×4
E ÷2 B

Finish number _____

2.

10
A ÷2
C −4 → D
×3
+8
E ×1 B

Finish number _____

3.

10
A +4
C −4 → D
+2
×3
E ÷2 B

Finish number _____

- **Now start at B with the number** $\boxed{10}$ **and return to B.**

4.

A +8
C −5 → D
+2
×4
E ÷2 **B**
10

Finish number _____

5.

A ÷2
C −4 → D
+2
+8
E ×1 **B**
10

Finish number _____

6.

A +4
C −4 → D
+2
×3
E ÷2 **B**
10

Finish number _____

- **Start with the number** $\boxed{10}$ **at C, then D, then E. Write the finish numbers.**

Now try this!

A +8
C −5 → **D**
+2
×4
E ÷2 B

A ÷2
C −4 → **D**
×3
+8
E ×1 B

A +4
C −4 → **D**
+2
×3
E ÷2 B

C to C	_____	_____	_____
D to D	_____	_____	_____
E to E	_____	_____	_____

Why do you think you get different finish numbers when you start with the same number at different places? Talk about it with a partner.

Teachers' note Ensure the children understand how the stars work: for example, for the first star begin at A with the number 10, then add 8 to reach B, then divide by 2 to reach C, and so on until multiplying by 4 to reach the finish number. As a further extension, invite the children to create their own stars, or to investigate other start numbers using the stars on this page.

Developing Numeracy
Mental Maths Year 5
© A & C BLACK

Notebook numbers

• **Complete these tables. Give your answers as decimals.**

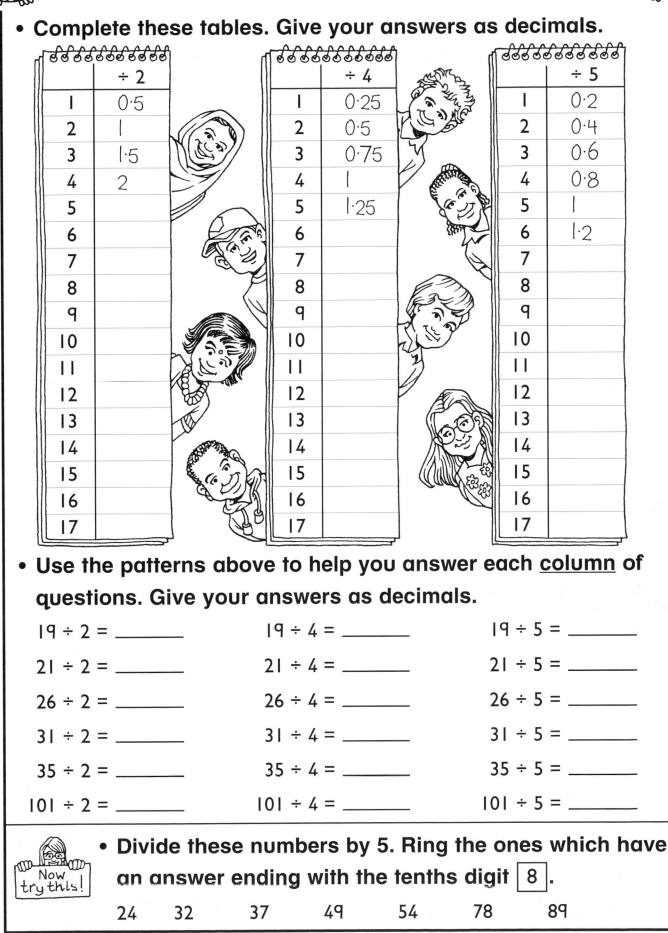

	÷ 2
1	0.5
2	1
3	1.5
4	2
5	
6	
7	
8	
9	
10	
11	
12	
13	
14	
15	
16	
17	

	÷ 4
1	0.25
2	0.5
3	0.75
4	1
5	1.25
6	
7	
8	
9	
10	
11	
12	
13	
14	
15	
16	
17	

	÷ 5
1	0.2
2	0.4
3	0.6
4	0.8
5	1
6	1.2
7	
8	
9	
10	
11	
12	
13	
14	
15	
16	
17	

• **Use the patterns above to help you answer each <u>column</u> of questions. Give your answers as decimals.**

$19 \div 2 =$ _____ $19 \div 4 =$ _____ $19 \div 5 =$ _____

$21 \div 2 =$ _____ $21 \div 4 =$ _____ $21 \div 5 =$ _____

$26 \div 2 =$ _____ $26 \div 4 =$ _____ $26 \div 5 =$ _____

$31 \div 2 =$ _____ $31 \div 4 =$ _____ $31 \div 5 =$ _____

$35 \div 2 =$ _____ $35 \div 4 =$ _____ $35 \div 5 =$ _____

$101 \div 2 =$ _____ $101 \div 4 =$ _____ $101 \div 5 =$ _____

Now try this! • **Divide these numbers by 5. Ring the ones which have an answer ending with the tenths digit 8 .**

24   32   37   49   54   78   89

Teachers' note The children should be familiar with remainders and have some understanding of tenths and hundredths before tackling this sheet. At the start of the lesson, revise fractions such as $\frac{1}{2}$, $\frac{1}{4}$, $\frac{1}{5}$, and ask the children to give these fractions as decimals.

Developing Numeracy Mental Maths Year 5 © A & C BLACK

Mental percentages

- **Use mental methods to calculate these percentages.**
  Check your answers using an equivalent calculation.

1.
```
        120
   50%  /    \  25%
  [60] ─── ÷ 2 ─── [  ]
```

2.
```
        240
   50%  /    \  25%
  [  ] ─── ÷ 2 ─── [  ]
```

3.
```
        180
   50%  /    \  5%
  [  ] ─── ÷ 10 ─── [  ]
```

4.
```
        360
   50%  /    \  5%
  [  ] ─── ÷ 10 ─── [  ]
```

5.
```
        220
   10%  /    \  30%
  [  ] ─── × 3 ─── [  ]
```

6.
```
        110
   10%  /    \  70%
  [  ] ─── × 7 ─── [  ]
```

7.
```
          400
   50%   25%   75%
  [  ] ÷2 [  ] ×3 [  ]
```

8.
```
          260
   50%   25%   75%
  [  ] ÷2 [  ] ×3 [  ]
```

9.
```
          320
   10%   5%   15%
  [  ] ÷2 [  ] ×3 [  ]
```

10.
```
          380
   10%   5%   15%
  [  ] ÷2 [  ] ×3 [  ]
```

- **Sketch diagrams to show how these percentages are linked.**

  Now try this!

  10%, 30% and 15%   10%, 1% and 7%   10%, 2% and 8%

  10%, 2% and 12%   10%, 1% and 99%   10%, 5% and 2·5%

Teachers' note At the start of the lesson, demonstrate how percentages of a number can be found mentally: for example, finding 50% by halving and finding 10% by dividing by 10. Discuss how other percentages can be found by multiplying or dividing these answers: for example, multiplying 10% by 3 to find 30%.

**Developing Numeracy
Mental Maths Year 5
© A & C BLACK**

Digital magic

A number is 'digitally magic' if it comes out of the machine with its digits in the reverse order.

12 → +9 → 21

The digits 1 and 2 are reversed. So 12 is digitally magic.

• **For each machine, tick the numbers that are digitally magic.**

1. 12 ✓   17
19   23
28   34
56   78
+9

2. 14   19
25   30
36   47
58   69
+27

3. 12   15
27   37
48   51
59   62
+36

4. 13   17
24   28
35   39
41   44
+54

5. 12   20
16   36
24   44
28   48
÷4   ×7

6. 15   30
25   55
35   50
45   40
÷5   ×6

7. 12   14
16   18
20   22
24   40
÷2   ×9

8. 12   15
18   21
24   27
30   33
÷3   ×8

Now try this!

• **Write a two-digit number between 11 and 20. Then reverse its digits. Subtract the smaller number from the larger.**

• **Try every number between 11 and 20.**

What do you notice about your answers?

• **Now try other two-digit numbers.**

Teachers' note Encourage the children to use suitable mental methods to find the output numbers, and ask them to discuss their strategies with a partner. Invite the children to say anything they notice about the answers: for example, that they are all multiples of 12 or that their digits are consecutive numbers.

**Developing Numeracy
Mental Maths Year 5
© A & C BLACK**

Currency crisis

At the bank, all the computers have crashed. The staff need to change money into different currencies.

- Use mental methods to convert the amounts.

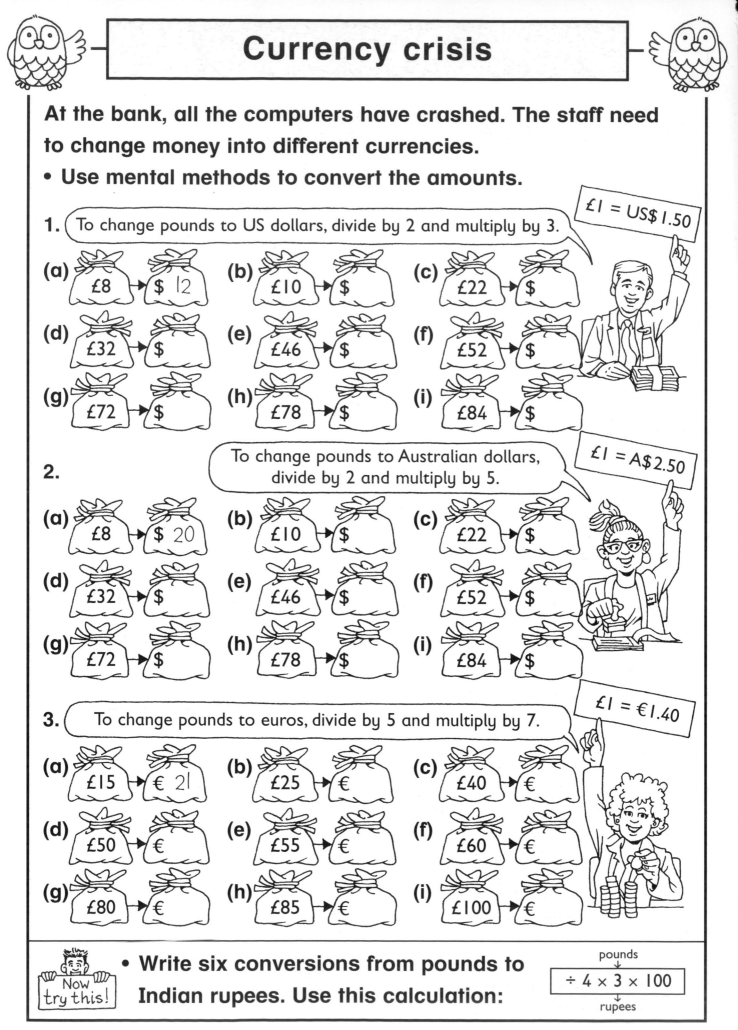

1. To change pounds to US dollars, divide by 2 and multiply by 3.

£1 = US$ 1.50

(a) £8 → $ 12   (b) £10 → $ (c) £22 → $

(d) £32 → $ (e) £46 → $ (f) £52 → $

(g) £72 → $ (h) £78 → $ (i) £84 → $

2. To change pounds to Australian dollars, divide by 2 and multiply by 5.

£1 = A$ 2.50

(a) £8 → $ 20   (b) £10 → $ (c) £22 → $

(d) £32 → $ (e) £46 → $ (f) £52 → $

(g) £72 → $ (h) £78 → $ (i) £84 → $

3. To change pounds to euros, divide by 5 and multiply by 7.

£1 = € 1.40

(a) £15 → € 21   (b) £25 → €   (c) £40 → €

(d) £50 → €   (e) £55 → €   (f) £60 → €

(g) £80 → €   (h) £85 → €   (i) £100 → €

Now try this!

- Write six conversions from pounds to Indian rupees. Use this calculation:

pounds
↓
÷ 4 × 3 × 100
↓
rupees

Teachers' note As a further extension, or as a means of checking their answers, the children could be asked to convert money from another currency to pounds, using inverse calculations (for example, divide by 3 and multiply by 2 to convert US dollars to pounds).

**Developing Numeracy
Mental Maths Year 5
© A & C BLACK**

Answers

p6

1. + 100	**2.** − 10	**3.** + 100 000
4. + 1000	**5.** + 10 000	**6.** − 1000
7. − 100 000	**8.** + 1	**9.** − 1
10. − 0·1	**11.** + 10	**12.** + 0·1

Now try this!

42 994	43 004	43 014	43 024
6·82	6·92	7·02	7·12
83 759	82 759	81 759	80 759
78 541	88 541	98 541	108 541
3·54	3·53	3·52	3·51

p7

1. 32	**2.** 4300	**3.** 76
4. 570	**5.** 210	**6.** 980
7. 960	**8.** 4·6	**9.** 6·2

Mira

Now try this!

£4.50	6·2 m	£11.30
£67.60	3·25 m	£17.58

p9

1. 3°C
2. 10°C
3. ⁻4°C
4. 2°C
5. 8°C → − 5 → + 1 → − 9 → − 2 → + 3 = ⁻4°C

Now try this!
10 → − 7 → − 4 → + 5 → − 3 → − 9 → ⁻8

p10

1.

	+4 →			
(23)	27	31	35	39
26	(30)	34	38	42
29	33	(37)	41	45
32	36	40	(44)	48
35	39	43	47	(51)

+3 ↓   (23) (30) (37) (44) (51)

7 is added each time

2.

	+8 →			
(15)	23	31	39	47
20	(28)	36	44	52
25	33	(41)	49	57
30	38	46	(54)	62
35	43	51	59	(67)

+5 ↓   (15) (28) (41) (54) (67)

13 is added each time

3.

	+4 →			
(53)	57	61	65	69
60	(64)	68	72	76
67	71	(75)	79	83
74	78	82	(86)	90
81	85	89	93	(97)

+7 ↓   (53) (64) (75) (86) (97)

11 is added each time

4.

	+6 →			
(27)	33	39	45	51
34	(40)	46	52	58
41	47	(53)	59	65
48	54	60	(66)	72
55	61	67	73	(79)

+7 ↓   (27) (40) (53) (66) (79)

13 is added each time

p13

1. 2, 4, 5	Other factors: 1, 10, 20
2. 2, 3, 4, 6, 8	Other factors: 1, 12, 24
3. 2, 4, 7	Other factors: 1, 14, 28
4. 3, 5, 9	Other factors: 1, 15, 45
5. 2, 3, 6, 7	Other factors: 1, 14, 21, 42
6. 2, 5	Other factors: 1, 10, 25, 50
7. 7	Other factors: 1, 49
8. 2, 4, 8	Other factors: 1, 16, 32

Now try this!

36	Other factors: 1, 12, 18, 36
56	Other factors: 1, 14, 28, 56

p15

1. pentagon	**2.** quadrilateral	**3.** hexagon
4. triangle	**5.** rectangle	**6.** square
7. quadrilateral	**8.** quadrilateral	**9.** octagon

p16

Now try this!
150 g + 450 g + 400 g
250 g + 350 g + 400 g
250 g + 300 g + 450 g
100 g + 250 g + 300 g + 350 g
100 g + 150 g + 300 g + 450 g

p17

1. 8	**2.** 11	**3.** 17
4. 5	**5.** 9	**6.** 14
7. 13	**8.** 18	**9.** 23
10. 13	**11.** 19	**12.** 22

p18

1. 359	**2.** 589	**3.** 688
4. 485	**5.** 583	**6.** 699
7. 591	**8.** 793	**9.** 892
10. 500	**11.** 660	**12.** 722
13. 827	**14.** 923	**15.** 923

Now try this!

896	682	239
794	318	937
709	904	601

p20

Now try this!

1. 0	**2.** 27	**3.** 14	**4.** 36
5. 158	**6.** 132	**7.** 108	**8.** 90

p21

Totals	Animals
20	PANDA
23	BEAR
24	CAMEL
28	TIGER
30	ZEBRA

p22

1. Totals always equal 210.
2. Totals always equal 290.

Now try this!
270   310   340
This is the sum of the numbers around the grid each time.

p23

Now try this!
59 at centre; 51, 52, 53 and 54 at corners; 55, 56, 57, 58 at
remaining places.

p24

All the differences are 44.

Now try this!
Other differences occur, e.g. difference might be 34 or 56 if one
digit is 0 or 1, and could be 24 or 156 if 2 digits are 0 or 1.

p25

1. 9·2	**2.** 8·8	**3.** 7·2
4. 12·0	**5.** 8·7	**6.** 12·3
7. 4·5	**8.** 6·4	**9.** 3·9
10. 2·7	**11.** 2·5	**12.** 1·4
13. 0·85	**14.** 0·89	**15.** 0·46
16. 0·81	**17.** 1·23	**18.** 1·18
19. 0·77	**20.** 0·35	**21.** 0·39
22. 0·34	**23.** 0·15	**24.** 0·27

ATLANTIC

p26

1. 51

2. 79	**3.** 56
4. 149	**5.** 204
6. 58	**7.** 100
8. 249	**9.** 300
10. 108	**11.** 242

Now try this!
7·9 cm   30 cm

p 27

1. 230	10	**2.** 150	90	**3.** 340	40
240	220	240	60	380	300
460	20	300	180	680	80
480	440	480	120	760	600
920	40	600	360		
960	880	960	240		

4. 400	40	**5.** 280	40	**6.** 260	160
440	360	320	240	420	100
800	80	560	80	520	320
880	720	640	480	840	200
				1040	640
				1680	400

Alternate numbers double down the columns.

p 29

1. 45	**2.** 64	**3.** 126
4. 90	**5.** 135	**6.** 195
7. 187	**8.** 165	**9.** 256

Now try this!
8, 4, 2, 6 in clockwise or anti-clockwise order

p 33

£800	£1400	£2200
£2600	£3200	£3800
£4800		£7000
£9400		£10 200
£11 200	£13 400	£14 600
£17 000	£17 800	£19 200

Now try this!
6900   8800   9500   9400   9700

p 34

1440 → 45
1040 → 65
1920 → 15
1760 → 55
1280 → 5
1360 → 85
19 200 → 75
16 000 → 125
12 800 → 25

Now try this!
575 → 18 400
115 → 1840
225 → 14 400
525 → 16 800
1125 → 144 000
105 → 1680

p 35

1. 80	**2.** 700
3. 300	**4.** 90
5. 240	**6.** 600
7. 60	**8.** 630
9. 2400	**10.** 210

Now try this!

180	350	540
70	1900	490
1300	450	630
810	270	2700
3000	460	3800

p 37

1. 18
21
24
30

2. 25	**3.** 16
30	21
45	25
55	28

4. 13	**5.** 20
26	13
42	42
50	50

6. 40	**7.** 15
140	50
200	30
330	110

Now try this!
one-fifth of 400, one-twentieth of 1600,
one-fortieth of 3200, one-eighth of 640

p 38

2 × 3

2 × 2 × 3	2 × 7
3 × 5	2 × 3 × 3
3 × 7	5 × 5
2 × 3 × 7	3 × 3 × 5

90
210
270
180
150
630
315
540
810
525

p 39

1. 247	260	273
2. 304	320	336
3. 342	360	378
4. 570	600	630
5. 418	440	462
6. 323	340	357
7. 266	280	294
8. 361	380	399
9. 684	720	756

Now try this!

	209	608	988
627	836	513	722

p 40

Possible calculations include:

47 × 3 = 141	47 × 4 = 188*	47 × 5 = 235	36 × 3 = 108
36 × 4 = 144	36 × 5 = 180	62 × 3 = 186	62 × 4 = 248
62 × 5 = 310	54 × 3 = 162	54 × 4 = 216	54 × 5 = 270
73 × 3 = 219	73 × 4 = 292	73 × 5 = 365	

Now try this!

216	196	192	204
208	189	203	198*

*closest to 200

p 42

1. 24	**2.** 11	**3.** 15
4. 16	**5.** 8	**6.** 13

Now try this!

18	13	14
23	15	16
21	15	15

p 43

Now try this!

24   49   54   89

p 44

1. 60	30		**2.** 120	60	
3. 90	9		**4.** 180	18	
5. 22	66		**6.** 11	77	
7. 200	100	300	**8.** 130	65	195
9. 32	16	48	**10.** 38	19	57

p 45

The following numbers should be ticked:

1. 12, 23, 34, 56, 78	**2.** 14, 25, 36, 47, 58, 69
3. 15, 37, 48, 59	**4.** 17, 28, 39
5. 12, 36, 24, 48	**6.** 45
7. 18	**8.** 27

Now try this!
Answers are all multiples of 9.

p 46

1.

(a) $12	**(b)** $15	**(c)** $33
(d) $48	**(e)** $69	**(f)** $78
(g) $108	**(h)** $117	**(i)** $126

2.

(a) $20	**(b)** $25	**(c)** $55
(d) $80	**(e)** $115	**(f)** $130
(g) $180	**(h)** $195	**(i)** $210

3.

(a) €21	**(b)** €35	**(c)** €56
(d) €70	**(e)** €77	**(f)** €84
(g) €112	**(h)** €119	**(i)** €140